Cooperation Among Animals

Oxford Series in Ecology and Evolution
Edited by Robert M. May and Paul H. Harvey

Cooperation Among Animals

An Evolutionary Perspective

LEE ALAN DUGATKIN

Department of Biology
University of Louisville
Louisville, Kentucky

New York   Oxford
OXFORD UNIVERSITY PRESS
1997

Oxford University Press

Oxford   New York
Athens   Auckland   Bangkok   Bogota   Bombay   Buenos Aires
Calcutta   Cape Town   Dar es Salaam   Delhi   Florence   Hong Kong
Istanbul   Karachi   Kuala Lumpur   Madras   Madrid   Melbourne
Mexico City   Nairobi   Paris   Singapore   Taipei   Tokyo   Toronto

and associated companies in
Berlin   Ibadan

Library of Congress Cataloging-in-Publication Data
Dugatkin, Lee Alan, 1962–
Cooperation among animals : an evolutionary perspective / by Lee Alan Dugatkin.
p. cm.—(Oxford series in ecology and evolution)
Includes bibliographical references and index.
ISBN 0-19-508621-X; 0-19-508622-8 (pbk.)
1. Animal behavior.   2. Cooperativeness.   I. Title.   II. Series.
QL751.D746   1997
591.51—dc20   96-18864

9   8   7   6   5   4   3   2   1

Printed in the United State of America
on acid-free paper

*This book is dedicated to
the four most important people in the world to me:
my loving wife, Dana; my precious little boy, Aaron;
and my supportive parents, Harry and Marilyn.*

Preface

When undertaking this project, I knew that I had my work cut out for me. Despite the fact that the format of this book is straightforward, the subject material—cooperation—is not. In fact, there were at least two ways that this book could have been written. The first would have partitioned the book primarily from a theoretical perspective. This approach would have entailed an introductory section followed by chapters on "Byproduct Mutualism and Cooperation," "Reciprocity and Cooperation," "Group Selection and Cooperation," and "Kin Selection and Cooperation." The second way of dividing up the material would have been primarily taxonomic—that is, an introduction followed by chapters like "Cooperation in Taxa A," "Cooperation in Taxa B," and so on.

In the end, after weighing the pros and cons and talking with colleagues about each, I opted for a hybrid approach that slants a bit toward the second of these formats. Following an opening chapter on the long, illustrious history of cooperative behavior, the second chapter reviews theoretical work on the evolution of cooperation, including some recent work developed by my colleague, Michael Mesterton-Gibbons, and myself. Chapters 3 to 7 proceed to examine cooperation in fish, birds, nonprimate mammals, nonhuman primates, and eusocial insects. Within each of these chapters, whenever possible, examples are placed within the theoretical framework established in Chapter 2. Chapter 8 outlines the direction in which I hope the field will move in the future.

To even attempt to place every putative example of cooperative behavior within a single theoretical framework is beyond the scope of any single book. Furthermore, I make no pretense about even attempting to cover all studies of cooperation. As such, on numerous occasions throughout this book, I refer the reader to reviews and other appropriate literature on a specific topic. Many of these topics—for example, cooperative breeding in birds—have been instrumental in the development of behavioral ecology as a legitimate subdiscipline in biology, and I certainly mean no disrespect by not devoting space to them here. In fact, the opposite is true. I think that it would be an injustice to tackle such important questions in one section of a single chapter. No doubt some people will feel that a book on the evolution of cooperation simply must cover some topics that I have not and I apologize in advance to those who feel slighted that their work has not been covered herein.

The decision regarding which studies to include and which not to include in this book was clearly somewhat subjective and, in some cases, personal. It could hardly be otherwise when dealing with such a complex idea as cooperation. I'm sure that the method I used to select which studies to highlight will differ from that of others, and I cannot always say for certain why a particular study was included or excluded. Let me at least, however, try to give some sense of the perilous path I took in deciding such matters.

To begin with, as mentioned earlier, some topics were simply too vast to cover. When I first proposed this book to Oxford University Press, Bob May, co-editor of the new series in Ecology and Evolution, made it clear that he was not interested in a book that simply re-hashed "classic" studies in cooperation—the sorts of studies that have already worked their way into the textbooks. I have kept this statement in mind; and while I, of course, review the "classics," they are only a small fraction of the cases covered. Whether a study is well cited or not, I always try to tie it back to some theory on the evolution of cooperation. For example, if a putative case of cooperation is directly interpretable within a theoretical framework, but has received little attention in the literature to date, that in itself makes it more likely to get coverage in this book. Unfortunately, many cases of cooperation do not neatly fall into one category or another. With respect to such studies, they were chosen when I believed that, in principle, they could be deciphered—that is, experiments could be constructed to help us fit this behavior into some sort of theoretical framework—the overriding theme being that the reader be provided with as broad an overview of the evolution of cooperation as possible. Again, subjectivity and personal biases are inevitable in such a process.

Whenever possible, within each chapter, I tried to address examples of cooperation that: (1) fell into different categories of cooperation—that is, byproduct mutualism, reciprocity, group selection, kin selection, or some combination of these, (2) covered different contexts in which cooperation took place—for example, foraging, anti-predator behavior, and so on, and (3) spanned a number of different species within a particular taxon. This was not always easy, and no doubt others might have approached the task with a different set of criteria. Last, I tried to choose cases that I thought best captured the essence of cooperative behavior, or showed the potential to do so in the future. I readily admit that this was the most subjective element in my decision-making process, but it is arguably the most important as well.

With the exception of Chapter 2, there are very few equations in the text, and even those in Chapter 2 should be interpretable to readers with the equivalent of college-level algebra. I, with the help of my colleague Michael Mesterton-Gibbons, have made a conscious effort to write the text surrounding equations in such a way that even if one were forced to skip the math (and I would encourage everyone to at least give it a crack before moving on), the message makes its way through to the reader.

I would be remiss if I did not take a moment to mention early on why I do not have a chapter on cooperation in humans. I thought a great deal about writing such a chapter and vacillated, weighing the pros and cons. In the end, I decided against such a chapter. Once I delved deeply into the human cooperation literature, I realized that there are hundreds, if not thousands of experiments on cooperation in humans. But, this, in and of itself, would not be a sufficient cause to omit human studies from this book. Rather, the major problem I encountered was that almost none of these studies is presented in the light of evolutionary theory (sad, but true) and so one would need to sift through all the literature and then reformulate it all in the context of evolutionary ideas. In addition, cooperation in humans has been studied in such a wide variety of contexts that it would likely be hard to make any generalizations across such studies, even if they could be presented in an evolutionary framework. Finally, it is often very unclear exactly what experimenters even meant by "cooperation" in a particular human study.

This book was written with advanced undergraduates, graduate students, and primary researchers in mind as the target audience. While I imagine that my audience will be primarily biologists, I sincerely hope that psychologists, anthropologists, and political scientists will consider this book for their classes (or book shelves) as well. These disciplines have had a profound impact on the study of cooperative behavior, but have often (though not always) ignored the evolutionary implications of their own work. If this book makes individuals in the above-mentioned disciplines realize that the evolutionary approach has much to offer, it will have performed a great service.

Acknowledgments

Often researchers have difficulty pinpointing when they became interested in a particular problem. This is a dilemma I do not share. I know exactly when my interest in cooperative behavior was kindled—when I first starting working with Jerram L. Brown. Jerry was my Master's advisor, and he suggested that I might read the literature on cooperation, particularly some recent work by Manfred Milinski, and think about that as a possible topic for my Master's degree. As always, this was a very insightful suggestion by Jerry, and I am forever in his debt for it. Without Jerry's help and advice in the early stages of my career, there is no doubt that I would be doing something very different with my life these days.

I had the great fortune of going from one fantastic mentor to another. After working with Jerry Brown, I began my Ph.D. work on the evolution of cooperation as a student of David Sloan Wilson. David immediately treated me both as student and colleague, and I have learned so much from him that I do not even know where to start. Regardless of what I do, I will always be David's student, and he will always be my mentor (even though he constantly tells me to stop saying that).

Much of the theoretical work in this book is a result of collaborative work with Michael Mesterton-Gibbons, a mathematical biologist at Florida State University. The cooperator's dilemma game outlined in Chapter 2 was first created by Mike, and I owe him a great debt of gratitude for that, among many other contributions he has made to this book. My work with Mike always makes me rethink fundamental questions surrounding the evolution of cooperation, and he is a cherished colleague indeed.

I have had the privilege and honor of discussing (usually debating) questions surrounding the evolution of cooperation with Jane Brockmann, Linnda Caporael, Tim Clutton-Brock, Phil Crowley, Richard Connor, Luc-Alain Giraldeau, Jean-Guy Godin, William D. Hamilton, Alasdair Houston, Felecity Huntingford, John Lazarus, Steve Lima, Anne Magurran, Manfred Milinski, Ronald Noë, Martin Nowak, Craig Packer, Joel Peck, Kern Reeve, Beren Robinson, Steve Rissing, Craig Sargent, Paul Sherman, David Stephens, John Maynard Smith, Andy Sih, and David Westneat. I'd also like to thank Dr. W. C. Allee, who despite not having been with us since his death in 1955 (see Schmidt, 1957; Banks, 1985 for biographies), has had a profound impact on my studies of cooperation. Dr. Allee's path-breaking work on animal cooperation lead to the publication of *The Social Life of Animals* (1931), which was reprinted in 1951 under the title *Cooperation Among Animals*. It is in Dr. Allee's memory that I have chosen *Cooperation Among Animals: An Evolutionary Perspective* as the title of this book.

Asking any colleague to read a book manuscript for you is tantamount to sentencing them to a short stint in prison, and so I am particularly grateful to Michael Alfieri, Jerram Brown, Phil Crowley, Michael Mesterton-Gibbons, Joel Peck, and David Sloan Wilson for reading all or parts of this book, and providing much-needed suggestions.

Writing this book has been a challenge in many ways, one that could not have been met without the help of many, many colleagues and friends. So many that I am bound to forget to list a few, and ask forgiveness for those who I inadvertently omit. The following individuals were kind enough to send me preprints, reprints, and manuscripts at various stages of development: J. Alcock, S. Altmann, K. Aoki, K. Armitage, M. Bekoff, N. Blurton-Jones, B. Borstnik, R. Boyd, M. Brewer, G. Burghardt, L. Caporael, D. Cheney, S. Creel, E. Curio, N. Davies, R. Dawes, R. Dawkins, F. de Waal, S. Emlen, I. Eshel, R. Fagen, E. Fischer, W. Foster, R. Gadagkar, M. Gadgil, J. Galef, T. Getty, P. Grant, M. Gross, B. Hart, C. Hemelrijk, B. Holldobler, J. Hoogland, A. Houston, S. Hrdy, Y. Iwasa, I. Jamieson, R. Jeanne, H. Kaplan, S. Komorita, C. Kullberg, J. Lazarus, E. Leigh, J. Ligon, S. Lima, M. Lombardo, H. Markl, M. Masters, M. McGuire, R. Michod, M. Milinski, R. Mumme, S. Nee, E. Nevo, R. Noë, M. Nowak, J. Orbell, J. Peck, O. Pellmyr, P. Petit, A. Pusey, D. Queller, E. Ranta, F. Ratneiks, H. Reeve, H. Reyer, P. Richerson, S. Riechert, G. Robinson, M. Rosenzweig, A. Rypstra, D. Scheel, P. Schmid-Hempel, T. Seeley, G. Sella, R. Seyfarth, P. Sherman, K. Sigmund, E. Sober, P. Stacey, D. Stephens, J. Strassmann, K. Sullivan, M. Taborsky, P. Taylor, R. Taylor, T. Valone, S. Vehrencamp, M. Wade, S. Wasser, J. Werren, G. Wilkinson, and G. Woolfenden.

I would also like to thank Chris Boesch, Stan Braude, Dorothy Cheney, Scott Creel, Frans de Waal, Marc Elgar, Ben Hart, Charlotte Hemelrijk, Thomas Kunz, Kern Reeve, Paul Sherman, Michael Taborsky, and AnthroPhoto Inc. for sending photos/slides/diagrams of various animals cooperating with one another.

Robert May and Paul Harvey have put together a wonderful series in ecology and evolution for Oxford University Press, and I am honored to have this book as part of it. Bob and Paul have been supportive all along the way and have done a great job of balancing the conflicting demands of (1) dealing with authors who are quite attached to what they have written and (2) making certain that the book accomplishes the goals for the Ecology and Evolution series. Kirk Jensen, Senior Science Editor for Oxford University Press (New York), has also been an invaluable resource in helping put this book together. I cannot thank Kirk, Bob, and Paul enough for their advice, patience, and encouragement.

If not for the help of my wonderful wife Dana, and the encouragement of my entire family, this book would never have been possible. Dana has been too supportive for words, done too much proofreading to discuss in public, and for someone with no training in the field of biology, has constantly amazed me with her insightful comments on all aspects of this book. Although my two-year-old son Aaron did not take much of an active role in helping me write this book, looking into his precious little eyes always reminded me to place this book, and the work leading up to it, in perspective and to remember what is really important in life.

Lastly, I would like to acknowledge my dear friend Henry Bloom for knowing me well enough to push me in the right direction and my brother David for seeing to it that I could reasonably defend why anyone in their right mind would head in that direction.

Louisville, Kentucky   L.A.D.
Spring 1996

Contents

Cooperation Among Animals

1

Historical perspectives on cooperative behavior

"Begin at the beginning," the King said very gravely, "and go on till you come to the end: then stop." (The Red King in Lewis Carroll's *Through the Looking Glass*)

Man is by nature a social creature: an individual who is unsocial naturally and not accidentally is either beneath our notice or more than human. Society is something in nature that precedes the individual. Anyone who either cannot lead the common life or is so self-sufficient as not to need to and therefore does not partake of society is either a beast or a god. (Aristotle, 328 B.C.)

Hereby it is manifest, that during the time men live without a common power to keep them all in awe, they are in a condition called war; and such a war, is of every man against every man. (Hobbes, 1651)

Act only on such a maxim through which you can at the same time will that it should become universal law. (Kant's Categorical Imperative)

1.1 Introduction

Evidence that great minds have been contemplating a question for millennia does not, in and of itself, make that subject interesting—but it certainly hints that it is. The introductory quotes by Aristotle, Hobbes, and Kant are meant to illustrate the long and illustrious history surrounding the study of cooperation. Cooperative behavior, in one form or another, has been an integral part of virtually all religions and has tickled the fancy of brilliant minds in philosophy, political science, economics, anthropology, psychology, and evolutionary biology throughout recorded history (and no doubt before this).

I'll begin with a review of early philosophical positions on cooperation and subsequently examine the history of this subject within evolutionary biology in an attempt to pave the way for a fuller understanding of the more technical material that follows. Any attempt at such a review must, however, be incomplete, as a thorough examination of cooperation across time and disciplines would require a series of books. As such, only a rough sketch, which highlights the ideas of major thinkers in the field, is given. Many of these ideas, however, are developed in considerably more detail in the chapters that follow.

1.2 Early philosophical ponderings on cooperation

The questions of good and evil, of man's tendency to be in a state of peace or war, and man's proclivity to cooperate or to cheat when the option presents itself have been a preoccupation of philosophers from at least the time of Aristotle. Some, like John Locke, have argued that although man may enter into a state of war, he is by nature cooperative:

> And here we have the plain difference between the state of nature and the state of war, which however some men have confounded, are as distant as a state of peace, mutual assistance and preservation: and a state of enmity, malice, violence and mutual destruction are from one another. (Locke, 1690; in Barker, 1962, p. 13)

Not all philosophers accepted the rosy view of man's nature portrayed by Locke and Aristotle (opening quote of this chapter). In the *Perke Avoth* (*Words of Our Fathers*), a 4th-century Hebrew book of law, members of society are handed this rather ominous warning: "Pray for the welfare of the government, since but for the fear thereof men would swallow each other alive" (*Perke Avoth,* ch. 3).

This statement regarding man's nature was greatly expanded by the British philosopher/political scientist Thomas Hobbes in his classic treatise *Leviathan* (1651). In *Leviathan,* Hobbes was consumed with mankind's inclination to noncooperative behavior and with the laws and codes a commonwealth needed to devise to overcome these noncooperative tendencies. Hobbes believed that the fundamental law of nature (as he referred to it) can be summed up as "every man has a right to everything: even to one another's body." If this is the case, argued Hobbes, then any sort of cooperation within society must be enforced by a sovereign in a commonwealth, because "covenants, without the sword, are but words, and of no strength to secure man at all" (Hobbes, 1651, p. 129). Without such commonwealth covenants, man's life is "solitary, poor, nasty, brutish and short" (p. 100). Despite his pessimistic view of cooperation in human society, Hobbes did not believe that cooperation was unnatural to all animals. In fact, in *Leviathan* he notes how cooperation occurs naturally in social insects. Hobbes, however, goes further, and uses the example of cooperation in social insects as a case study of why man is *not* cooperative in his natural state:

> It is true, that certain living creatures, as bees and ants, live sociably one with another, which are therefore by Aristotle numbered amongst political creatures; and yet have no other direction; than their particular judgments and appetites; nor speech, whereby one of them can signify to another, what he thinks expedient for the common benefit; and therefore some man may desire to know, why mankind cannot do the same. To which I answer:
>
> First, that men are continually in competition for honor and dignity,

which these creatures are not; and consequently amongst men there ariseth on that ground, envy and hatred, and finally war; but amongst these not.

Secondly, that amongst these creatures, the common good differeth not from the private and being by nature inclined to their private, they procure thereby the common benefit. But man, whose every joy consisteth in comparing himself with other men, can relish nothing but what is eminent.

Thirdly, that these creatures, having not, as man, the use of reason, do not see, nor think they see any fault, in the administration of their common business; where as amongst men, there are very many, that thinks themselves wiser, or abler to govern the public, better than the rest; and these strive to reform and innovate, one this way, another that way; and thereby bring into distraction and civil war.

Fourthly, that these creatures, though they have some use of voice, in making knownst to each other their desires, and other affections; yet they want that art of words, by which some men can represent to others that which is good, in the likeness of evil; and the evil in the likeness of good; and augment, or diminish the apparent greatness of good and evil; discontenting men, and troubling their peace at their pleasure.

Fifthly, irrational creatures cannot distinguish between injury and damage; and therefore as long as they be at ease, they are not offended with their fellows; whereas man is then most troublesome, when he is most at ease; for then it is that he loves to shew his wisdom, and control the actions of them that governth the commonwealth.

Lastly, the agreement of these creatures is natural; that of men is by covenant only, which is artificial; and therefore it is no wonder if there be somewhat else required, besides covenant, to make their agreement constant and lasting; which is common power, to keep them in awe, and to direct actions to the common benefit. (Hobbes, 1651, ch. 17, pp. 111–12)

1.3 Early evolutionary biology and the study of cooperation

It is hard to find an interesting topic in modern behavioral and evolutionary ecology that cannot, in one sense or another, be traced back to the work of Charles Robert Darwin. Cooperation is no exception—Darwin was very worried about cooperative behavior and what it implied about his theory of natural selection. After all, how could natural selection operate on cooperative behavior when it appears that donors pay a cost, yet benefits are accrued by other group members? For example, In *The Origin of Species* (1859), Darwin spent considerable space outlining the various cooperative and altruistic tendencies that almost always define the social instincts. He went on to say that such behavior was "a special difficulty, which at first appeared to me insuperable, and actually fatal to my whole theory." In a characteristic flash of brilliance, however, Darwin resolved this paradox by outlining inclusive fitness theory more than 100 years before Hamilton (1963, 1964a,b). Like Hamilton after him, Darwin recognized that if natural selection operated at the level of the social insect *colony*, many features of the social insects, including their cooperative and altruistic tendencies, were readily understandable. For example, when discussing how natural selection might produce sterile worker castes with dramatically different morphologies from those destined for reproduction, Darwin noted: "The difficulty, though appearing insuperable, is lessened, or as I believe disappears, when it is remembered that selection may be applied to the family, as well as to the individual, and thus may gain the desired end" (1859, p. 204).

Darwin's interest in cooperative behavior was by no means limited to the social insects. For example, he wrote at length about allogrooming, sentinel calls, and cooperation in birds and mammals, and made the rather bold claim that "the most common mutual service in the higher animals is to warn one another of danger by means of the united senses of all" (Darwin, 1871, p. 474). Furthermore, in *The Descent of Man* (1871), he extended his analysis to humans and expressed fascination in the bravery displayed by primitive tribe members during warfare. How could such cooperative tendencies evolve, when individuals who display them are more likely to be killed in warfare than those who do not? To answer this question, Darwin was forced, once again, to argue that selection at some level other than the individual was the cause of such cooperative behaviors:

> When two tribes of primeval man, living in the same country came into competition, if (other circumstances being equal), the one tribe included a great number of courageous, sympathetic and faithful members, who were always ready to warn each other of danger and aid and defend each other, this tribe would succeed better and conquer the other. (1871, p. 498)

Darwin's "bulldog," Thomas Henry Huxley, also displayed a keen interest in cooperative behavior, or more accurately, the lack of cooperation in the animal kingdom. While Darwin (1859) acknowledged that the struggle for existence is often metaphorical, insofar as it is often a struggle against the environment, Huxley, in *The Struggle for Existence and Its Bearing upon Man* (1888), took a more extreme view and believed "From the point of view of the moralist, the animal world is on about the same level as the gladiator's show. The creatures are fairly well treated, and set to fight; whereby the strongest, the swiftest and the cunningest live to fight another day. The spectator has no need to turn his thumb down, as no quarter is given."

In a manner that would have made Hobbes proud, Huxley then extended this view to primitive man:

> [T]he weakest and the stupidest went to the wall, while the toughest and the shrewdest, those who were best fitted to cope with their circumstances, but not the best in any other way, survived. Life was a continuous free fight, and beyond the limited and temporary relations of the family, the Hobbesian war of each against all was the normal state of existence.

Whether Huxley's gladiator view of life is accurate or not, it caused very strong reactions from those who believed that cooperation was the norm in nature, rather than the exception. For example, A. R. Wallace (1891), the co-discoverer of the theory of natural selection, took strong exception to Huxley's portrayal of animal life. Wallace believed that whatever struggle took place led to animals developing behaviors, including cooperative behavior, to alleviate the misery outlined by Huxley. In *Darwinism* (1891), Wallace argued that "On the whole, then, we conclude that the popular idea of the struggle for existence entailing misery and pain on the animal world is the very reverse of the truth. What it really brings about, is, the maximum of life and of enjoyment of life with the minimum of suffering. . . ." (p. 40).

Wallace took this view to even further extremes in the context of human behavior. While a hard-line "Darwinist" on the role that natural selection played in shaping all traits of animals, Wallace took a sharp U-turn when it came to human behavior, which he believed could be explained only by the presence of a supernatural deity. To Wallace, "the whole purpose, the only raison d'etre of the world . . . was the development of the human spirit in association with the human body" (1891, p. 477).

Peter Kropotkin, a Russian prince, was perhaps the most prominent and pro-lific opponent of Huxley's "gladiator" view of animal behavior. Kropotkin be-lieved that it was an injustice to Darwin and his theory to intimate that coopera-tion was rare. He wrote a series of letters-to-the-editor in response to Huxley's 1888 paper. In the end, these letters developed into his famous *Mutual Aid—A Factor of Evolution* (1908)*,* in which he argued that despite the fact that the works of Darwin show him to be acutely aware of the large role cooperation played in animal social life, Darwin's advocates often intentionally ignored this aspect of his theory.

Kropotkin is a fascinating character in the history of evolutionary biology. In addition to being a member of the Russian royal family, a founder of the anarchist movement, and a world-renowned geologist, he also happened to be an ardent natural historian. In his extensive, worldwide travels, Kropotkin saw what he took to be cooperative behavior among animals at every turn: "[I]n all the scenes of animal life which passed before my eyes, I saw mutual aid and mutual support carried on to an extent which made me suspect in it a feature of the greatest importance for the maintenance of life, the preservation of each species and its further evolution" (1908, p. 18).

Kropotkin's emphasis on mutual aid in animals was undoubtedly tied to his libertarian views of social justice (see the preface to Kropotkin, 1908), and his rhetoric often combined these political views with Darwin's theory of natural selection: "if we resort to an indirect test and ask Nature: 'Who are the fittest: those that are continually at war with each other or those who support one another?,' we at once see those animals which acquire mutual aid are undoubt-edly the fittest" (1908, p. 30).

Kropotkin, however, did not deny that competition played a role in animal life. Rather, he argued that although competition often existed for food, it was predation and abiotic factors that were responsible for the vast majority of deaths in the animal world, and mutual aid was the best response to such fac-tors. While not an experimental biologist, Kropotkin provided vivid details of cooperation in the context of nest defense in bees and ants, cooperative hunting in birds and wolves, sentry posts in parrots, and the building and maintenance of prairie-dog towns.

It is arguable that of all the books on cooperation written by biologists, Kropotkin's *Mutual Aid* had the most profound affect on biologists, social scientists, and laymen alike. Montagu (1952), in *Darwin, Competition and Cooperation,* a book that echoed many of Kropotkin's ideas (and is in fact dedicated "To the memory of Peter Kropotkin..."), notes:

> Kropotkin's book is now a classic—which means that few people read it and that it is now out of print. Yet no book in the whole realm of evolutionary theory is more readable or more important, for it is *Mutual Aid* which provides the first thoroughly documented demonstration of the importance of cooperation as a factor in evolution. Kropotkin's book, one may be sure, is destined for a revival, and the influence it has already had is likely to increase many fold with the years. (p. 42)

After (or in some cases, just preceding) Kropotkin's classic, and before experimental work on cooperation began (see below), a number of books that touched on cooperation were published. Of particular note are Espinas's *Des Societies Animals* (1878), Romanes's *Mental Evolution in Animals* (1895) and *Animal Intelligence* (1898), Alverdes's *Social Life in the Animal World* (1927), Reinheimer's *Evolution by Cooperation, a Study in Bioeconomics* (1913) and *Symbiosis, a Sociophysiological Study of Evolution* (1920), and Patten's *The Grand Strategy of Evolution* (1920) and *The Passing of Phantoms: A Study of Evolutionary Psychology and Morals* (1925). Yet none of these had the effect that *Mutual Aid* had on the scientific and non-scientific communities.

1.4 Experimental work on cooperation: Warder Clyde Allee and "The Chicago School" of animal behavior

W. C. Allee, cofounder of the "Chicago School" of animal social behavior (Mitman, 1992), believed that "the ordinary thoughtful person is not aware that the tendency toward a struggle for existence is balanced by the strong influence of the cooperative urge." Despite all his work in the area, however, Allee apparently never explicitly defined cooperative behavior, perhaps because he believed that its definition was so clear. From his examples, however, Allee clearly believed cooperative behavior to be an all-encompassing phenomenon and included a fair share of examples that most modern-day behavioral ecologists would not consider to be cooperative. Allee felt that there existed a continuum of cooperative behaviors that manifested itself in creatures from bacteria to man, and he used such terms as "proto-cooperation," "automatic cooperation," and "unconscious cooperation" when describing his work on very simple creatures such as bacteria. It is perhaps fair to say that what Allee's continuum truly measured was not cooperation per se, but rather behaviors that promoted group living and enhanced individual productivity within such groups.

Allee argued that the first stage of cooperation in simple organisms like bacteria and worms consisted of merely tolerating the presence of other individuals. Once groups formed, individuals in such groups began accruing benefits

(Banks, 1985). Documenting these advantages became Allee's passion and preserved his name in ecology texts forever. The "Allee Effect"—namely, that undercrowding, not just overcrowding, can have disadvantages—is known to even first-year students of ecology, evolution, and behavior.

In *Animal Aggregations* (1931) and *The Social Life of Animals* (1938), Allee mustered his evidence that cooperation is a fundamental feature of animal social life, even among the simplest creatures. A sample of the studies that Allee cited in *The Social Life of Animals* includes work showing that: (1) goldfish and daphnia survive longer in toxic environments as group size increases; (2) Planaria survival in ultraviolet light is a function of group size; (3) per capita rate of growth in bacteria is a function of group size; (4) goldfish grow faster in groups, because they regurgitate food that is available to groupmates; (5) amphibians regenerate their tails faster when living in groups; (6) time from hatching to fledging in colonial birds is reduced in large colonies; and (7) in certain contexts, minnows and goldfish learn tasks faster when living in groups.

In fact, Allee believed that cooperation could be demonstrated in virtually all taxa:

> Actually there are in the scientific literature good cases of mass protection for almost all the animals shown in these charts; and where exact information was lacking, for example in the rotifers, this was a result of lack of interest in conducting experiments on this point with these animals. Eventually, it turned out as predicted, that mass protection for rotifers could be demonstrated. (1938, p. 57)

This view did not go unchallenged, however. For example, Raymond Pearl, the famous population biologist, took exception with Allee's views on the ubiquity of cooperative behavior and asked:

> Does the mere fact that n animals survive longer than $n + m$ or $n - m$ animals in an unfavorable solution in itself demonstrate that there is any element of cooperation, automatic or otherwise? . . . Is not further evidence, and of a qualitatively different sort necessary to prove such a condition? (1939, p. 306)

Aside from documenting the ubiquity of cooperation, Allee's other (related) passion was studying dominance hierarchies. Mitman (1992) shows nicely how Allee's interest in dominance hierarchies was intimately tied to his Quaker lifestyle, in that Allee wished to demonstrate that dominance hierarchies were in fact a type of cooperative system in and of themselves. For example, Allee and his students studied hierarchy formation in chickens and argued that "the peck-order type of social organization of flocks of hens may serve to build a cooperative social unit better fitted to compete or to co-operate with other flocks at the group-level than are socially unorganized groups" (Guhl et al., 1945).

In *The Social Life of Animals,* Allee further extended this claim to human cooperation:

> Much can be said for an established order of dominance and subordination, whether within groups of non-human animals or among nations. There is

growing evidence that with hens, again as an example, well organized flocks, in which each individual knows and is fairly resigned to its particular social status, thrive better and produce more eggs than do similar flocks that are in a constant state of organizational turmoil. Similarly, among nations, relative quiet exists when the international order of dominance is fairly established and generally accepted. (p. 204)

Such "naive" group-selectionist explanations (Wilson and Sober, 1989) of social behavior were not uncommon at the Chicago School. Mitman (1992) in fact has argued that the rise of population biology as a subdiscipline meriting interest was intimately tied with such group-selectionist views. Fledgling population biologists needed as much support as they could muster that the population was a distinct entity, and group-selection views helped provide such evidence.

The study of cooperation in the Chicago School was by no means the exclusive domain of Allee (Mitman, 1992). Alfred Emerson, evolutionary biologist, ethologist, and the premier termite expert of his day, was also quite interested in cooperative behavior, particularly in the context of insect societies as superorganisms (Emerson, 1939, 1946, 1960; also see Wilson and Michener, 1982, for a full listing of Emerson's work). While Allee was interested in cooperation as an end, to Emerson cooperation was a means for creating homeostasis within insect colonies and increasing group-level productivity. Mitman (1992) discusses this as well as Emerson and Allee's different views on the role of kinship in the evolution of cooperation in his very readable book, *The State of Nature: Ecology, Community and American Social Thought, 1900–1950.*

1.5 "Modern" approaches to the study of cooperative behavior

Needless to say, it is always difficult to delineate where "modern" work on any subject, including cooperation, begins and "older" work ends, as one naturally flows into the other. Nonetheless, one way to create such a dichotomy is to consider modern work on cooperation as that undertaken after the existence of a sound theoretical framework was established. The work that I have discussed until this point had no formal framework within which to generate and test hypotheses regarding the evolution of cooperative behavior. Such a framework sprang into existence in 1963 and 1964, with W. D. Hamilton's seminal papers on inclusive fitness. In these papers, which many believe mark not only the beginning of the modern study of cooperation and altruism, but also the start of modern behavioral ecology and sociobiology, Hamilton argued that altruistic and cooperative behaviors are more likely to evolve among relatives, because such individuals share more genes that are identical by descent than do individuals randomly drawn from the population at large:

In the hope that it may provide a useful summary we therefore hazard the following generalized unrigourous statement of the main principle that has emerged from the model. *The social behavior of a species evolves in such a way that in each distinct behavior-evoking situation the individual will seem to value his neighbors' fitness against his own according the coefficients of relationship appropriate to that situation.* (1964a, p. 19; emphasis in original)

(See Chapter 2 for more details on Hamilton's inclusive theory and its relation to various other mechanisms that may select for cooperative behavior.)

Even to those unable to grind their way through the math, Hamilton's theory made a clear prediction: cooperation should be relatively more common among related than unrelated individuals. The impact of inclusive fitness theory on studies of cooperative behavior cannot be overstated. Hamilton's (1964) work, in conjunction with J. L. Brown's 1970 and 1974 essays, spurred the now booming communal/cooperative breeding "industry" (Brown, 1994). And while work on social and eusocial insects had been active already for over a century, Hamilton's (1964) theory, in connection with Wilson's *Insect Societies* (1971), provided those who studied these fascinating creatures with a theoretical framework for generating specific predictions (see Chapter 7 for more on cooperation in social insects).

Hamilton's work did not, however, solve the riddle of altruism and cooperation among *unrelated* individuals—what E. O. Wilson (1975) referred to as "the central theoretical problem of sociobiology." While inclusive fitness theory served as a framework from which to study cooperation among related individuals, no comparable theoretical foundations had been laid for examining the evolution of cooperation among unrelated individuals. Trivers's (1971) seminal article is rightly regarded as the first conceptual/theoretical attempt at cracking the nut of altruism among nonrelatives. Trivers (1971) argued that genes for altruistic acts may be selected if individuals differentially allocate such altruism to those who have been altruistic themselves—so-called "reciprocal altruism." In this same article, there is a rarely cited paragraph in which Trivers framed the question of reciprocal altruism in the language of game theory and the prisoner's dilemma:

> The relationship between two individuals repeatedly exposed to the symmetrical reciprocal situations is exactly analogous to the what game theorists call the Prisoner's Dilemma . . . W. D. Hamilton has shown that {the above treatment of} reciprocal altruism can be re-formulated concisely in terms of game theory as follows. . . . (p. 39)

The prisoner's dilemma (Fig. 1.1), first created by Flood and Dresher of the Rand Corporation and popularized by Von Neumann and Morgenstein (1953), takes its name from the following scenario: two suspects of a crime are interrogated by the police, while in separate rooms. Cooperation and defection are defined from the perspective of the suspects—to defect means to "squeal" and tell the authorities that the other suspect is guilty and to cooperate is the converse. The police have enough circumstantial evidence to put away both suspects for 1 year, even without a confession from either. Should each suspect "rat" on the other, however, both go to jail for 3 years. Finally, should only one suspect snitch on his partner, such "state's evidence" allows

Player 2

	Cooperate	Cheat (Defect)
Cooperate	R 1-year prison term	S 5-year prison term
Cheat (Defect)	T 0-year prison term	P 3-year prison term

Fig. 1.1. The prisoner's dilemma game. The game is defined by the inequalities $T > R > P > S$, and $2R > T + S$. Payoffs are shown for player 1.

the cheater to walk away a free man, but causes his partner to go to jail for 5 years. As can be seen in Figure 1.1, on any single play of the game, player 1 receives a higher payoff for cheating, regardless of what player 2 does ($T > R$ and $P > S$). The game is constructed such that we can switch the player 1 and 2 labels, and hence player 2 should also defect. The dilemma is that both players would have received a higher payoff if they both cooperated than if they both defected ($R > P$). (See Poundstone, 1992, for more on the history of the prisoner's dilemma game. Also see Maynard Smith and Price, 1973, for early work on the hawk-dove game, which shares many attributes with the prisoner's dilemma.)

The prisoner's dilemma game came into the spotlight again with Axelrod and Hamilton's (1981) paper on the evolution of cooperation among unrelated individuals. Axelrod (1980a,b), who had been interested in cooperation from a political scientist's perspective, wrote Richard Dawkins at Oxford University about work on evolutionary approaches to studying cooperation (Dawkins, 1989). Dawkins responded that the person Axelrod needed to speak with, W. D. Hamilton, was at Axelrod's own university (Michigan). In their subsequent collaboration, Axelrod and Hamilton (1981), using both analytical techniques and computer simulations, examined the success of an array of strategies in the iterated prisoner's dilemma game (see Chapter 2 for more on this). They searched for evolutionarily stable strategies (Maynard Smith, 1982) to a game in which players are paired up with some probability of encountering each other in future interactions. They argued that the approach developed differs from preceding work on the subject in three ways:

1) In a biological context, our model is novel in its probabilistic treatment of the possibility that two individuals may interact again . . . 2) Our analysis of the evolution of cooperation considers not just the final stability of a given strategy, but also the initial viability in an environment dominated by noncooperating individuals, as well as the robustness of a strategy in a variegated environment composed of other individuals using a variety of more or less sophisticated strategies. . . . 3) Our application includes behavioral interactions at the microbial level. (Axelrod and Hamilton, 1981, p. 1391)

What they found was that if the probability of meeting a given partner in the future was above some critical threshold, then in addition to the success of a simple strategy of "always defect" (ALLD), a conditionally cooperative strategy called "Tit for Tat" (TFT) was a robust solution to the iterated prisoner's dilemma. TFT instructs a player to cooperate on the initial encounter with a partner and to subsequently copy its partner's last move. Axelrod (1984) hypothesized that TFT's success is attributable to its three defining characteristics: (1) "Niceness"—TFT is never the first to defect; (2) Swift "retaliation"—TFT immediately defects on a defecting partner; and (3) "Forgiving"—TFT remembers only one move back in time. As such, TFT forgives prior defection, if a partner is currently cooperating (i.e., it does not hold grudges). Essentially, TFT succeeds because it segregates behavior at the phenotypic level—that is, it matches moves of cooperation with cooperation, and defection with defection (Michod and Sanderson, 1982).

Axelrod and Hamilton's work spurred a plethora of theoretical articles on the evolution of cooperation. Much of the early work generated results that were concordant with those of Axelrod and Hamilton, but more recent studies have called into question: (1) the "robustness" of TFT—it appears that strict TFT is not nearly as invincible as Axelrod and Hamilton (1981) believed (TFT-like strategies, however, appear to be quite robust: Nowak and Sigmund, 1992, 1993), and (2) whether one should expect to see pure TFT or pure defection as a solution to the iterated prisoner's dilemma. Recent work has shown that TFT and defecting strategies can coexist at equilibrium (Peck and Feldman, 1986; Boyd and Lorberbaum, 1987; Farrell and Ware, 1989; Dugatkin and Wilson, 1991). I discuss these newer developments in later chapters. Somewhat surprisingly, a boom in empirical studies of reciprocity among animals and the use of TFT did not immediately follow Axelrod and Hamilton's (1981) article.

1.6 Onward and upward

In the past, my colleagues and I have argued that, in addition to reciprocity, there are at least three other 'categories of' cooperative behavior: kin selection, group selection, and byproduct mutualism (Dugatkin et al., 1992; Mesterton-Gibbons and Dugatkin, 1992). I will now move on to outlining these four categories in considerable detail. From there, I examine examples of cooperation in various taxa and how they fit into the four mechanisms mentioned above.

2

Theoretical perspectives on the evolution of cooperation

I am a firm believer that without speculation there is no good and original observation. (Charles Darwin in a letter to A. R. Wallace, 1867)

2.1 Introduction

As we saw in the opening chapter, evolutionary biologists and others have had a longstanding interest in cooperative behavior. Up to this point, however, the word "cooperate" has been tossed around rather loosely and no formal definition has yet been provided. This was partially intentional, as it seems difficult to provide such a definition in the context of (briefly) summarizing 2300 years of thought on what is in fact a broadly defined word, at least in everyday usage. In fact, even within the field of animal behavior and behavioral ecology, attempts to describe terms like cooperation and altruism sometimes produce almost oxymoronic phrases such as "self-interested refusal to be spiteful" (Grafen, 1984), "quasi-altruistic selfishness" (West-Eberhard, 1975) and "joint stock individualism" (Kropotkin, 1908; see Wilson and Dugatkin, 1992, for more on these terms and the semantic confusion surrounding them).

To eliminate any further ambiguities, I open this chapter with a definition. *Cooperation* is an outcome that—despite *potential* relative costs to the individual—is "good" in some appropriate sense for the members of a group, and whose achievement requires collective action. But the phrase "to cooperate" can be confusing, as it has two common usages. To cooperate can mean either: (1) to achieve cooperation—something the group does, or (2) to behave cooperatively, that is, to behave in such a manner that renders the cooperation possible (something the individual does), even though the cooperation may not actually be realized unless other group members also behave cooperatively. Here, to cooperate will always mean to behave cooperatively.

There has been no lack of major theoretical and conceptual papers which have attempted to tackle the issue of why animals cooperate, or fail to cooperate, with one another. Theoreticians, in particular, have a penchant for building mathematical (usually game theoretical) models to address this question. Any attempt to document, let alone review all the models of cooperative behavior constructed by evolutionary biologists, political scientists, psychologists, and economists would not only be difficult, but futile, because new models are published at such a rate as to make any such list immediately outdated (but see Axelrod and Dion, 1987; Axelrod and D'Ambrosio, 1994, for a partial bibliography on the evolution of cooperation via reciprocity). Within evolutionary biology, however, models of cooperation can be divided into four basic categories: (1) cooperation via kin selection, (2) cooperation via group selection, (3) cooperation via reciprocity, and (4) cooperation via byproduct mutualism (Dugatkin et al., 1992; Mesterton-Gibbons and Dugatkin, 1992; Dugatkin and Mesterton-Gibbons, 1995). After discussing these various approaches, I will move on to outline a single framework that incorporates them.

2.2   Inclusive fitness

In what are now undoubtedly the most cited pair of articles in the behavioral and evolutionary ecology literature, W. D. Hamilton (1964a,b) tackled the question of the evolution of altruism and cooperation among kin in his "The Genetical Theory of Social Behaviour I and II." It was in these papers that Hamilton, building on ideas introduced by Haldane (1932), introduced the concept of inclusive fitness.

In many ways the beauty of Hamilton's theory is its "distilled" simplicity. That is, despite some nasty mathematics, the basic idea conveyed makes intuitive sense. The "kernel" of inclusive fitness models is that they supplement "classical" models by considering the effect of a gene not only on the individual that bears it, but on others as well, most importantly those sharing genes that are identical by descent (i.e., kin). Fortunately for the mathematically fainthearted, the equations in these models can be simplified to what has now become referred to as "Hamilton's rule," namely that a gene increases in frequency whenever

$$b/c > 1/r \tag{2.1}$$

where b = the benefit associated with the trait the gene codes for, c = the cost accrued by manifestation of this trait, and r = Wright's coefficient of relatedness (in Hamilton's original work, he represented b/c by the parameter k, and so the inequality was $k > 1/r$). This rule is often presented for the case of diploid full siblings as $b/c > 2$. Grafen (1984) has argued that for statistical testing purposes, Hamilton's rule is better rearranged and presented as $rb - c > 0$, as the variance in r may be calculated from data, but the variance of the ratio b/c is more complicated.

Hamilton's inclusive theory has had a profound impact on the sorts of studies undertaken by behavioral and evolutionary ecologists. Although work had been going on in the field of cooperative breeding in birds for decades (Skutch, 1935, 1987; see Brown, 1994, for a review), Hamilton's papers, in conjunction with Brown's (1970) empirical work on Mexican jays, caused a surge in studies in cooperative and communal breeding that is still in full force today (see

Brown, 1987, for a graphical depiction of the relationship between Hamilton's theory and work in communal breeding). Furthermore, the whole field of kin recognition was in no small part a result of Hamilton's theory. And last, in the most general sense, behavioral ecologists and evolutionary biologists now accept inclusive fitness as *the* framework in which to think about the process of natural selection.

The impact of inclusive fitness theory on communal/cooperative breeding was even greater than it would have been as a result of J. L. Brown's reformulation of Hamilton's rule. One restriction of Hamilton's approach is that inclusive fitness was cast within generations, but much of the early field work generated by these models was looking at cooperation across generations. Furthermore, field workers found the *b* and *c* terms of Hamilton's model too vague to study in nature. Brown's (1975) "offspring rule" addressed both these problems. The offspring rule can be written

$$G \cdot r_H > L \cdot rB \qquad (2.2)$$

where G is the indirect fitness gain to helping, L is the direct fitness loss as a result of not breeding, r_H is the relatedness of the helper to the young it helps raise, and r_b is the relatedness of parent to offspring. Brown (1975, 1994) observed that the offspring rule differs from Hamilton's rule in that although one can be derived from the other, the offspring rule is designed to study altruism between individuals of different generations, while Hamilton's rule is designed to examine altruism between individuals of the same generation. Brown (1987) notes:

> [T]he essential difference can be seen to lie in r_B, which is the relatedness of parent to offspring, and is normally 1/2. The difference arises from the fact that in Hamilton's rule gains and losses are measured in "fitness" of individuals in the same generation, while in the offspring rule, they are measured in offspring, hence the name. (p. 53)

Although at first glance Hamilton's rule and the offspring rule appear easy enough to apply, many researchers have fallen into the trap of "double accounting" in which the affects of a trait are counted in both directions, rather than just one. For more on how and how not to measure inclusive fitness, I refer the reader to Brown (1975, 1987), West-Eberhard, (1975), Charnov (1977), Wade (1980), Michod and Hamilton (1980), Seger (1981), Michod (1982), Grafen (1984, 1985), Creel (1990), Queller (1992) Lucas et. al (1996), and Queller (1996). It is also important to note that while "Hamilton's rule" in its simple form can be quite powerful, when one considers inclusive fitness from a population genetic standpoint, things can get quite complicated, particularly when natural selection is strong (e.g., see Levitt, 1975; Cavelli-Sforza and Feldman, 1978; Uyenoyama and Feldman, 1981; Uyenoyama 1984; Toro et al., 1982; Matessi and Karlin, 1984; Mueller and Feldman, 1988).

2.3   Within- and between-group selection

A related framework for studying the evolution of cooperation is that of "trait-group" selection. Before discussing the means by which trait group selection explains the evolution of cooperation, it is necessary to have a slight sortie back in time in order to distinguish between old (sometimes referred to as "naive") group selection and newer trait group models.

Although most evolutionary biologists think of group selection as a rather contentious issue, it is important to realize that until the 1960s, evolutionary biologists and ecologists did not spend much effort arguing about the level at which natural selection acts. A peaceful coexistence existed between group and individual selectionists, in part because the levels-of-selection question was simply not recognized as an issue. As such, the rampant group selection of Kropotkin (1908), Allee (1943), and Emerson (1960) stood side by side with work on selection at the level of the individual. Unfortunately, aside from Haldane (1932) and Wright's (1945) brief "island model" of group selection, virtually no theoretical work addressed the levels-of-selection question until the 1960s and 1970s.

This peaceful coexistence vanished in the late 1950s and 1960s with the works of Williams and Williams (1957), Wynne-Edwards (1962), Hamilton (1963, 1964a,b, 1975), Brown (1966), Maynard Smith (1964), and Williams (1966, 1971). In the longest, most vehement, but least rigorous argument to date, Wynne-Edwards (1962) presented his case for group selection as the major force controlling population size, particularly in birds. Despite the fact that Wynne-Edwards provided no theoretical underpinnings to his ideas, he nonetheless felt free to make broad, sweeping, unsubstantiated remarks on the efficacy of group selection (see Wynne-Edwards, 1986, for a toned-down version of these arguments).

By 1967, group selection was dealt an apparently devastating set of blows by the works of Hamilton (1964a,b) and Williams (1966). Williams argued that virtually all cases of purported group selection could be understood within the framework of classical individual selection. Invoking Occam's razor, he proposed that we need not invoke group selection as an explanation in such cases, and made some very strong statements to this effect. For example, Williams states, rather boldly,

> "With some minor qualifications . . . , it can be said that there is no escape from the conclusion that natural selection, as portrayed in elementary texts and in most of the technical contributions of population geneticists, can only produce adaptations for the genetic survival of individuals. (1966, pp. 7–8)

Williams's case against group selection rests on two premises. First, although selection above the level of the individual is theoretically possible, in practice such selection will be weak because of the relative speed of within-versus between-population selection. Second, virtually no evidence exists that cannot be understood using the individual selectionist paradigm. The effect of

Williams's prose and Hamilton's mathematical models on the status of group selection was a blow that group selectionists are still recovering from today. D. S. Wilson (1983, p. 159) noted, "For the next decade group selection rivaled Lamarckianism as the most thoroughly repudiated idea in evolutionary biology." Surprisingly, at the same time that Dawkins (1976) was developing his selfish gene theory as an extension of Williams's (1966) ideas, a "new group selection" school arose in the mid-1970s. While Dawkins (1976) was presenting his reductionist view that genes, but not groups (or for that matter, individuals), qualified as "replicators" (i.e., enduring units of self-interest), the new group selection school was being spearheaded by D. S. Wilson's trait-group models (1975, 1976, 1977, 1980) and the empirical and theoretical work of Wade (1977, 1978, 1979).

Three critical features separated Wilson and Wade's views of group selection from that of their predecessors. First and foremost, they provided detailed genetic models which partitioned variance into within- and between-group components (a procedure developed in Williams and Williams, 1957). Such models allowed detailed predictions of the circumstances favoring the evolution of individual versus group beneficial traits (also see Price, 1972; Eshel, 1972; Boorman and Levitt, 1973a,b; Levin and Kilmer, 1974; Gadgil, 1975; Cohen and Eshel, 1976; Matessi and Jayakar, 1976). Second, the definition of a group was no longer confined to a reproductively isolated deme. Wilson (1975) introduced the term "trait group" and defined it as a population "within which every individual feels the effect of every other individual" (Wilson, 1980, p. 22). Importantly, trait groups are seen as embedded within a larger interbreeding population, and thus models of trait-group evolution are sometimes called intrademic selection models, to distinguish them from the interpopulational or interdemic models such as those implied in the arguments of Wynne-Edwards (1962). Third, Wynne-Edward's style group selection relied on a very special mechanism—extinction caused by overexploitation of a resource. Trait-group models rely on superior group level production, whatever the mechanism, and are hence much more general. The difference between "old" group selection and trait-group selection is shown diagrammatically in Figure 2.1.

The "trait-group" approach to the evolution of cooperation is quite straightforward. Cooperation can evolve even when it has a relative cost to the individual performing it, if this *within-group cost* is offset by some *between group-benefit*, such that *cooperative groups are more productive than selfish groups* (for more on this see Wilson and Sober, 1994). For such group-level benefits to be manifest, groups must differ in the frequency of cooperators within them, and groups must be able to "export" the productivity associated with cooperation. For example, following Wilson (1990a), imagine a metapopulation of social insect colonies. In the primitive condition there are no "specialized foragers"—those who take the often high risk associated with foraging, but share food with nest mates. How could such a cooperative "specialized forager" evolve? If we denote the unspecialized strategy as S (selfish), and the specialized strategy as A (altruistic), then:

Assume that queens in all-S colonies forage equally, while A forages disproportionately often in colonies of one A and $(N - 1)S$. Let $x < 1$ equal A's probability of survival relative to S in colonies all-S and let $y > 1$ equal A's effect on her own colonies' chances of prevailing over the other colonies. Globally, the fitness of A will exceed the fitness of S when $xy > 1$. Thus, within-group selection against the trait (represented by x) can be arbitrarily low, as long as the between-group selection for the trait (represented by y) is sufficiently high. (Wilson, 1990, p. 139)

As we will see in Chapter 7, this example closely mimics what happens in a number of species of social insects.

Such trait group models are typically used to explain cooperation among unrelated individuals, but need not be restricted as such (see Dugatkin and Reeve, 1993, for a method of translating between "group" and "individual" selection).

2.4   Reciprocity and the prisoner's dilemma

Because a good deal of the work on the evolution of cooperation uses game theoretical models, and because the theory I describe later in this chapter employs this modeling technique, I will spend a fair share of time on this subject. The workhorse of game theory models examining reciprocity and the evolution of cooperation is undoubtedly the prisoner's dilemma (see Fig. 1.1). Although the subject of at least hundreds, if not thousands, of studies in the field of psychology (Colman, 1982; Caporeal et al., 1989), and despite its similarity to the hawk-dove game (Maynard Smith and Price, 1973), most evolutionary biologists were unaware of the prisoner's dilemma game before the collaborative work of Robert Axelrod, a political scientist, and W. D. Hamilton. Axelrod and Hamilton (1981) were interested in both the emergence and subsequent evolution of cooperation in a world in which defection was the primitive state.

To examine the evolution of cooperation in such a world, Axelrod and Hamilton (1981) used a two-prong attack. First, Axelrod ran computer tournaments in which participants submitted a strategy in the form of a computer program and strategies competed against one another in games that lasted 200 moves (see Axelrod, 1984, for details). Fourteen strategies were submitted (see Axelrod, 1984, for a list), and each strategy faced itself and all others (including an entry that was programmed to choose between cooperation and noncooperation in a random fashion) in a round-robin tournament. Tit for Tat (TFT), submitted by Anatol Rapoport, emerged as the winner of this tournament. Recall from Chapter 1 that TFT is a strategy that instructs players to cooperate when first meeting an opponent and to subsequently copy whatever that opponent does. Further recall that TFT has three characteristics that allow it to do well in such tournaments: TFT is nice (always starts off cooperatively), retaliatory (defects in response to defection), and forgiving (does not hold grudges—remembers only one move back).

Axelrod followed up the original tournament with a second that differed from the original in two ways. First, it contained a wider variety of submitted strategies (62 in all; see Axelrod, 1984, for a list), and second, Axelrod undertook what he referred to as an "ecological" tournament. In the ecological tour-

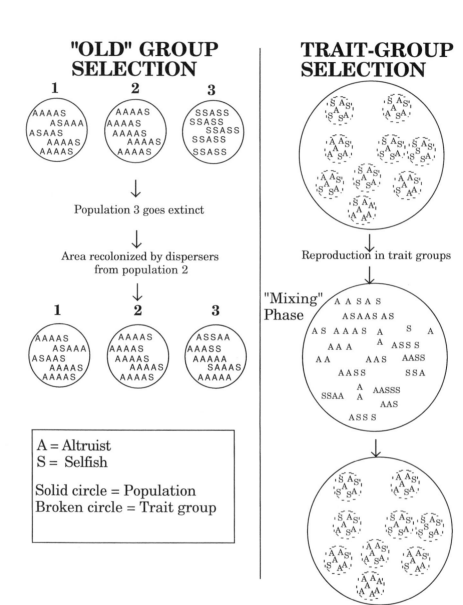

Fig. 2.1. In "old" group selection models, populations are reproductively isolated. The probability that a group goes extinct is proportional to the frequency of altruists (A) in it. Here, population 3 is assumed to go extinct as a result of the high frequency of selfish (S) types it contains. Population 2 colonists then recolonize this patch as a result of the high frequency of altruists within this unit. Altruism is unlikely to evolve under this scenario for two reasons. First, groups are reproductively isolated, and over time any group containing even a single S moves rapidly toward fixation of S. Second, groups with many altruists are unable to "export" their productivity, since extinction of moderately sized groups is assumed to be rare. In trait-group models, trait groups are embedded within a population and reproduction occurs within these groups. Again, selfish types increase within groups, but groups with many altruists outproduce those with few. After reproduction, trait groups dissolve and a global "mixing" phase occurs. The mix-

nament (which might better be called an "evolutionary" tournament), after the initial round-robin, a second "generation" is simulated in which strategies exist in proportion to their success in the original round-robin. This process is iterated over time, to determine what strategies would evolve in the long run. Once again, TFT emerged as the winner.

In addition to these computer tournaments, Axelrod and Hamilton (1981) examined whether TFT and the primitive ALL DEFECT strategies were evolutionarily stable, that is, whether they could resist invasion from mutants if they themselves were at a frequency close to 1. They began this task by proving that if ALLD or a strategy that alternates D and C (ALTDC) could not invade TFT, then no single pure strategy could invade. They then considered whether ALLD and or ALTDC could in fact invade TFT.

If we let w equal the probability of interacting with the same player on the next move of a game, then moves of the game are a geometric series and the expected number of interactions with a given opponent is equal to $1/(1 - w)$. So, for example, if $w = 0.9$, the expected number of interactions with a given partner is 10 [$1/(1 - .9)$]. This being the case, when TFT is close to fixation, virtually all TFT players meet other TFTs and their payoff is:

$$R + wR + w^2R + w^3R + \ldots \rightarrow R/(1 - w) \tag{2.3}$$

An ALLD mutant would have all its interactions with TFT, and its payoff would be:

$$T + wP + w^2P + w^3P + \ldots \rightarrow T + wP/(1 - w) \tag{2.4}$$

and so TFT can resist invasion from ALLD when

$$R/(1 - w) > T + wP/(1 - w) \tag{2.5}$$

Solving the inequality for w, ALLD fails to invade TFT when

$$w > (T - R)/(T - P) \tag{2.6}$$

Now, ALTDC gets a payoff:

$$T + wS + w^2T + w^3T \ldots \rightarrow (T + wS)/(1 - w^2) \tag{2.7}$$

when playing TFT, and thus TFT is resistant to invasion when:

Fig. 2.1 (*continued*)
ing phase may occur many times during an individual's life, and as such, altruistic groups are provided numerous opportunities to "export" their productivity. After the mixing phase, new trait groups are formed. Altruism evolves here when the increased productivity of groups with many altruists outweighs the within-group benefit to being selfish (after Dugatkin and Reeve, 1993).

$$R/(1 - w) > (T + wS)/(1 - w^2) \qquad (2.8)$$

Solving the inequality for w, ALTDC fails to invade TFT when:

$$w \geq (T - R)/(R - S) \qquad (2.9)$$

and hence Axelrod and Hamilton (1981) concluded that TFT is resistant to any invasion when:

$$w \geq \max[(T - R)/(T - P), (T - R)/(R - S)] \qquad (2.10)$$

Clearly, then, the success or failure of cooperation in the iterated prisoner's dilemma depends on the probability of future play with the same player, or what has been termed "the shadow of the future." Although TFT can resist invasion from ALLD and ALTDC, a strategy of "always cooperate" (ALLC) could "drift" into a population of TFT (since TFT would always play C against such a mutant) and eventually go to high frequencies. So, to be precise, what Axelrod and Hamilton demonstrated was that TFT is "collectively stable" rather than evolutionarily stable, the difference being that for collective stability a strategy must do *as well as, or better than* any mutant, but for evolutionary stability a strategy must do *better than* a mutant.

It is also important to note, as do Axelrod and Hamilton (1981), that ALLD is always resistant to invasion (at least in their analysis), regardless of the value of w. This, of course, poses a problem for the *initiation* of TFT. Axelrod and Hamilton (1981) suggest that an "initial clustering" of TFT might overcome this threshold problem:

> Suppose that a small group of individuals is using a strategy such as TIT FOR TAT and that a certain proportion, p, of the interactions of members of this cluster are with other members of the cluster. Then the average score obtained by the members of the cluster in playing TIT FOR TAT is $p[R/(1 - w)] + (1 - p)[S + wP/(1 - w)]$. If the members of the cluster provide a negligible proportion of the interactions for the other individuals, then the score attained by those using ALL D is still $P/(1 - w)$. When p and w are large enough a cluster of TFT can then become initially viable in an environment composed overwhelmingly of ALL D. (p. 1394)

It is worth noting that the evolution of TFT can also be couched in the language of within- and between-group selection. The simplest way to see this is to recall that in any TFT/ALLD pair, TFT always obtains fewer rewards than ALLD ($S + wP/(1 - w)$ versus $T + wP/(1 - w)$, respectively). It is only the existence of TFT/TFT pairs in which TFT gets a relatively high payoff ($R/(1 - w)$) and ALLD/ALLD pairs wherein ALLD does poorly ($P/(1 - w)$) that allows TFT to outdo ALLD. In other words, between-group (in this case, between-pair) selection for cooperation outweighs within-group selection against it (see Wilson, 1983; Dugatkin and Reeve, 1993; and Wilson and Sober,

1994, for more on this). However, since the behavioral mechanisms associated with reciprocity (memory, individual recognition, etc.) are interesting in and of themselves, and because not all strategies associated with reciprocity fit so nicely into the group selection framework, I believe that separating reciprocity from group selection will be a useful exercise.

One source of controversy in the analysis of the iterated prisoner's dilemma surrounds the question of which strategies to include in the game (in game theory terminology, what the strategy set should be). On the one hand, the set of all possible strategies is infinitely large; on the other hand, a strategy can be an evolutionarily stable strategy (ESS) only with respect to the other strategies, and not all strategies are guaranteed representation. For example, TFT is not an ESS with respect to any set that includes ALLC = (C, C, C, . . .). But if this set also includes ALLD = (D, D, D, . . .), then ALLC is "weakly dominated" by TFT (that is, ALLC is never better than TFT and sometimes worse); and if weakly dominated strategies are not represented, then TFT may be an ESS if w is sufficiently large. Thus, the practical importance of proofs that no strategy in the IPD can be an ESS (Boyd and Lorberbaum, 1987; Farrell and Ware, 1989; Lorberbaum, 1994) is simply that theoretical work is limited to finding necessary conditions (prerequisites), because sufficient conditions can never be obtained, as every strategy is invadable by some conceivable mix of mutant strategies. Such necessary conditions are strongly dependent on the pattern of interaction. For example, Mesterton-Gibbons (1992b) has shown that in the analyses of Axelrod (1980a,b) and Boyd and Lorberbaum (1987), an individual's opponent is drawn at random from an infinitely large population and retained for the duration of the iterated prisoner's dilemma; and that very different conditions occur when the population is finite. This analysis indicates that significant clustering may be necessary even to maintain cooperation, let alone initiate it; and so even if reciprocity can explain the persistence of cooperative behavior, we must invoke a different mechanism to explain its emergence.

May (1981) has noted that although Axelrod and Hamilton (1981) consider the "shadow of the future" in their parameter w, which discounts the probability of future *interaction,* their model does not does discount future *payoffs.* That is, the entries in the prisoner's dilemma game are constant and are not modified by some time discounting factor. May notes:

> In many biological applications, however, future gains (measured in terms of reproductive efforts or other coinage related to darwinian "fitness") may be worth much less than current gains. One offspring in the hand can literally be worth two in the future. I think Axelrod and Hamilton's analysis can be widened to include such discounting of future gains, at least in very simple cases, by formally redefining the probability of future encounters, w, to include an appropriate discount factor. If this is so, it adds a further requirement that must be satisfied before cooperation can evolve: not only must there be a sufficiently high probability that any two protagonists will encounter each other again, but also the gains from such future encounters must not be too heavily discounted. . . . (p. 292)

May's point has not received the attention it deserves from either empiricists or theoreticians.

Although Axelrod and Hamilton's (1981) article did not spur empirical studies of the IPD for about five years (see subsequent chapters), and even though there have been numerous challenges to the use of the prisoner's dilemma as a model for the evolution of reciprocity (Rothstein and Pirotti, 1987; Noë, 1990; Connor, 1995a), Axelrod and Hamilton's (1981) study immediately tantalized evolutionary theorists who have produced a plethora of articles following up their initial model. Modifications to the IPD include variations in population structure, number of players, number of strategies, relatedness of players, stochasticity of strategies, stochasticity of environment, amount of memory, possibility of individual recognition, norms, ostracism, mobility of players, and mistakes by players. These models are summarized in Table 2.1. It is not possible to discuss even a fraction of these papers in the text, but I trust that the table will give the reader a feel for their primary effects. In the following section, I sketch out two of these models. Although there is undeniably some subjectivity in which models were chosen, my purpose was to select a few models that I believe may have the greatest combined impact on empiricists and theoreticians.

2.4.1 Reciprocity and mistakes

Despite the old adage, death and taxes are not the only sure things in life. Another certainty is that animals, and perhaps especially non-human animals, make mistakes. One of the most fundamental problems with the reciprocity strategies associated with the early prisoner's dilemma game is that errors on the part of interactants were not taken into account (May, 1987). Of all the criticisms about game theory being removed from the "real world," this is perhaps the most reasonable, because errors can have very strong effects on the dynamics of the game. For example, suppose an individual defects by mistake. If it is interacting with players who use certain types of strategies (e.g., Tit for Tat), a single act of mistaken defection can cause an unlimited series of defect-defect interactions. Boyd (1989), following some initial work on this question by Selten and Hammerstein (1984) and economist R. Sugden (1986), addressed this issue by considering the possibility that an individual that intended to cooperate defects by mistake, and vice versa. Boyd's (1989) motivation for examining mistakes was to determine whether errors might allow some cooperative strategy to be an ESS to the iterated prisoner's dilemma. Boyd (1989) viewed this as an important issue, because his own earlier work had shown that no pure strategy was an ESS to the prisoner's dilemma (without mistakes) and he found this

> quite discouraging because it suggests that the nature of the strategies that will be able to persist in real populations depends on the distribution of rare phenotypic variants actually produced by mutation or environmental variation. (p. 47)

Table 2.1. Recent evolutionary models of reciprocity that modify Axelrod and Hamilton's (1981) work. Some studies fall under more than one heading but are organized by the primary modification made.

Modification	Result	Reference
N-person game	Increasing group size hinders the evolution of cooperation.	Joshi, 1987 Boyd & Richerson, 1988
	Increasing group size hinders the evolution of cooperation, but decreases the threshold frequency of TFT needed to invade ALLD.	Dugatkin, 1990 Dugatkin & Godin, 1992a
	Predator-prey interactions may result in selfish groups being more vigilant than cooperative groups.	Packer & Abrams, 1990
	Finds mixed ESSs to a "synergistic" N-person game where individuals "play the field."	Motro, 1991
Population structure	Within social networks, the evolution of TFT is "individual" specific. Players in the most dyads serve as a point of invasion for ALLD.	Pollock, 1988
	Reciprocity can resist invasion in viscous "lattice"-like population.	Pollock, 1989
	Group selection can favor cooperation.	Boyd & Richerson, 1990a
	Spatial prisoner's dilemmas allow very generous cooperative strategies.	Grim, 1995, 1996
	When players live in a spatially homogenous environment with densities varying in time and space, TFT and defect can coexist. This coexistence may be in a "pattern" (a spatially homogeneous stationary rate).	Hutson & Vickers, 1995
Stochastic environment	Kinship is not necessary for food sharing among communal breeders.	Caraco & Brown, 1986
	Variability in predation risk and value of food patches affects the evolution of food calling.	Newman & Caraco, 1989
	IPD as a Markov Chain. Solution contains "never play D after C." TFT is much weaker in a "probabilistic" world.	Nowak, 1990b Nowak & Sigmund, 1990
	ESSs to the IPD (and other games) may be inaccessible.	Nowak, 1990a

(continued)

Table 2.1. (continued)

Modification	Result	Reference
	Stochastic strategies can lead to cycles of reciprocity and defection.	Nowak & Sigmund, 1989
	"Pavlov" strategy outcompetes TFT.	Nowak & Sigmund, 1993
Definition of stability/ suites of strategies	Proves no pure strategy is an ESS to the IPD.	Boyd & Lorberbaum, 1987
	Proves no mixed strategy is an ESS to the IPD.	Farrell & Ware, 1989
	No strategy is an ESS to the IPD.	Lorberbaum, 1994
	Sets of strategies exist that do not lose against any other strategy.	Borstnik et al., 1990
	Generous TFT replaces TFT as best cooperative strategy.	Nowak & Sigmund, 1992
	Define types of evolutionary stability and their implications for the evolution of cooperation.	Bendor & Swistak, 1995
Indirect reciprocity	Networks of "indirect reciprocity" are unlikely to facilitate cooperation unless groups are fairly small.	Alexander, 1986 Boyd & Richerson, 1989
Genetic algorithms	Genetic algorithms come up with TFT-like strategies that are solutions to the IPD.	Axelrod, 1987
	Genetic algorithms are used to examine the relationship between individual recognition and cooperative behavior.	Crowley et al., 1995
Payoff structure	Payment function of next encounter is a random variable. Cooperation may benefit player in that it keeps opponent alive for next play.	Eshel & Weinshall, 1988
	If payoff structure of game changes as the frequency of strategies change, "helping" behavior can evolve from mutation frequency.	Peck & Feldman, 1986
	Considers dynamic game, where players can opt to play alone and where the game has finite number of moves. Cooperation can evolve even in the face of "end game" effects.	Lima, 1989
	Cooperative hunting is examined via a family of models. Cooperation likely when individual is unlikely to capture large prey alone. Individual differences in hunting ability select against TFT.	Packer & Rutton, 1988

Table 2.1. (continued)

Modification	Result	Reference
	Cooperator's dilemma allows cooperation to evolve via reciprocity, group selection, kin selection, or byproduct mutualism.	Dugatkin et al., 1992; Mesterton-Gibbons & Dugatkin, 1992
Kinship	TFT can invade "altruistic kin groups." With respect to other alleles, TFT is an outlaw gene.	Wilson & Dugatkin, 1991 Dugatkin et al., 1994
	Kinship promotes cooperation in small groups.	Aoki, 1983
	When the value of various associates differs, distant kin can easily be selected over close kin for partnerships.	Wasser, 1982
	Hamilton's rule holds for reciprocity, once "coefficients of synergism" are taken into account.	Queller, 1985
Mobility of players	"Roving" defectors hinder cooperation.	Dugatkin & Wilson, 1991 Dugatkin, 1992 Harpending & Sobus, 1987 Enquist & Leimar, 1993
	Movement of players is modeled as a diffusion process. If TFT players have significant mobility, they can spread in wavelike fashion through a population.	Ferriere & Michod, 1995
	Partner choice allows cooperation in yucca moth and fig-wasp system.	Bull & Rice, 1991
Social learning	Observer TFT strategy can invade TFT. "Reputation" evolves.	Pollock & Dugatkin, 1992
Encounter probabilities	Assortment favors the evolution of cooperation.	Eshel & Cavalli-Sforza, 1982 Michod & Sanderson, 1985 Toro & Silio, 1986
Encounter probabilities	"Friendship" can evolve when players can form and re-form partnerships.	Peck, 1993
	If probability of partner retention is low, population needs to be small for cooperation to evolve.	Mesterton-Gibbons, 1991, 1992b Mesterton-Gibbons & Childress, 1996

(*continued*)

Table 2.1. (continued)

Modification	Result	Reference
Ostracism	Ostracism favors cooperative behavior.	Hirshleifer & Rasmussen, 1989
Punishment of defectors	"Norms" help cooperation spread in a population.	Axelrod, 1986 Boyd & Richerson, 1992
Prior experience	If a player's choice of C or D is dependent on its own play on the prior move, cooperation may profit.	Feldman & Thomas, 1987 Thomas & Feldman, 1988
Memory/recognition	"Single partner" games are replaced by two parameters that measure encounters with known versus unknown players. Reciprocity can evolve from near zero frequency.	Brown et al., 1982
	Games with no individual recognition may qualify as a prisoner's dilemma under some ecological scenarios.	Mesterton-Gibbons, 1991
Alternating moves	Having opponents alternate moves in an IPD does not change qualitative predictions.	Boyd, 1988
	Generous TFT does well in a game where players alternate moves.	Nowak & Sigmund, 1994
	Challenges the notion that pure Generous TFT does well when players alternate moves.	Frean, 1994
Mistakes	When TFT makes errors, or there is a "cost of complexity" associated with playing TFT, it is no longer robust.	Hirshleifer & Martinez-Coll, 1988
	Mistakes by players allows an ESS to emerge in the IPD.	Boyd, 1989
	Mistakes hinder the evolution of reciprocal strategies like Pavlov.	Stephens et. al., 1996
Cultural transmission	Cultural transmission typically helps cooperation to evolve.	Boyd & Richerson, 1982 Boyd & Richerson, 1990b

Boyd (1989) considered a strategy called "Contrite TFT," created by Sugden (1986). Contrite TFT depends on the notion of "good standing," as follows:

> An individual is always in good standing on the first turn. It remains in good standing as long as it cooperates when CTFT specifies that it should cooperate. If an individual is not in good standing it can get back in good standing by cooperating on one turn. Then CTFT specifies that an individual should cooperate (i) if it is not in good standing, or (ii) if its opponent is in good standing; otherwise it should defect. To see why I call this contrite, consider the sequence of behaviors following a mistake by one member of a pair both playing CTFT. On turn t both players 1 and 2 intend to co-operate, but player 1 mistakenly defects. On turn $t + 1$, player 2 is good standing and player 1 is not. Thus player 1 co-operates, and player 2 defects. On turn $t + 2$, player 2 is still in good standing despite his defection on the previous turn because CTFT called for defection. Thus, player 1 absorbs the sucker's payoff as an act of contrition. (p. 51)

If a sufficiently high probability of future interaction exists, CTFT is evolutionarily stable. Given that animals make mistakes, CTFT allows for them to avoid the nasty situation in which they continue to alternate cooperating and defecting indefinitely, as they would if they were using simple TFT. In other words, the ability to repent for one's actions favors cooperative behavior.

2.4.2   "Reactive" strategies and the emergence of "Generous TFT" and "Pavlov"

A second critique of the computer simulations run by Axelrod (1980a, b) is that despite the fact that a large number of strategies competed against each other in his computer simulations, all such strategies were deterministic. That is, they were all of the form "if x happens, do y" (where x might be a single event or a series of events). Nowak and Sigmund (1992), among others, remedied this dearth of stochasticity by introducing "reactive" strategies. A reactive strategy is what Nowak and Sigmund (1992) refer to as a "triple" (y, p, q), where y is the probability of cooperating on the initial encounter with a stranger, and p and q are the probabilities of playing C after your opponent plays C or D, respectively. Such reactive strategies are even simpler when w is very large and the importance of first moves (and hence y) become negligible (hence strategies become "doubles" rather than "triples").

Nowak and Sigmund (1992) studied the evolution of cooperation among such reactive strategies, when w is very large (i.e., 1) and 100 strategies are pitted against one another. What they found was that for the basic "triples" paradigm, evolution proceeds toward a world of pure cheaters (strategies close to $p = 0$, $q = 0$). If, however, the game is slightly rigged in the sense that one TFT-like strategy (p close to 1, q close to 0) is put in place at the start of a simulation, evolution proceeds in a very different, but interesting manner. At first ALLD strategies start to spread, but after awhile TFT-like strategies make an unexpected comeback. However, the TFT-like strategy that put reciprocity back on the playing field is itself only a temporary player that sets the stage for the eventual winner, Generous Tit for Tat (GTFT). GTFT is a reactive strategy with a p value of nearly one and a q value of min$[1 - (T - R)/(R - S), (R - P)/(T - P)]$, which equals 1/3 for standard values prisoner's dilemma values of $T = 5$, $R = 3$, $P = 1$, $S = 0$. Nowak and Sigmund (1992) describe the rise of GTFT:

> In our simulations, TFT is almost specified by its police role: strategies that are not very close to TFT do not have its effect. But, an evolution twisted away from defection (and hence toward TFT) leads not to the prevalence of TFT, but toward more generosity. TFT's strictness is salutary for the community, but harms its own. TFT acts as a catalyzer. It is essential for starting the reaction towards cooperation. It needs to be present initially, only in a tiny amount; in the intermediate phase, its concentration is high; but in the end only a trace remains. (p. 252)

One comforting aspect of Nowak and Sigmund's (1992) results is that they are rather intuitive. That is, no animal, including man, *always* uses as strict a rule as TFT. Humans and, as we shall see in later chapters, other animals sometimes cooperate with those who have recently cheated on them, for whatever reason. Turning the other cheek is apparently not only the decent thing to do, it appears to be the direction in which natural selection moves in stochastic worlds.

The plot thickens, however, when reactive strategies take into account the actions of *both* players on the prior move (not just the move of an individual's partner, as in TFT and GTFT), and the dynamics of this game are examined over very long time scales (millions of generations). Nowak and Sigmund (1993a) considered the case in which reactive strategies were defined by four probabilities, p_1, p_2, p_3, and p_4, which measure the odds of cooperating after having received R, S, T, and P, respectively. In such a world a strategy called Pavlov, which cooperates if both players opt for the same behavior on the last move of the game (1, 0, 0, 1), outperforms both TFT and GTFT (see Wedekind and Milinski, 1996, for more on GTFT and Pavlov in humans). The dynamics of these simulations mimic punctuated equilibrium (Eldridge and Gould, 1972), in that populations find themselves in very long periods of stasis in which cooperation or defection is the behavior displayed by most players. These periods of stasis can end very dramatically with a complete reversal of fortune. In the end (4 million + generations in Nowak and Sigmund, 1993a), however, Pavlov and Pavlov-like strategies emerge victorious. Nowak and Sigmund (1993a) attribute Pavlov's success to two features:

> (1) an inadvertent mistake between players using TFT causes a drawn out battle: between Pavlovians, it causes one round of mutual defection followed by a return to joint cooperation. (2) Whereas TFT and GTFT can be invaded by drift by all-out cooperators (to the eventual profit of exploiters) Pavlov has no qualms in exploiting a sucker, once it has been discovered (after an accidental mistake) that it need not fear retaliation. (p. 56)

Whether Pavlov or something akin to it is used by non-humans remains to be seen.

2.5 Byproduct mutualism

Group selection, kin selection, and reciprocity have attracted the lion's share of attention in studies of cooperation. A fourth path to cooperation, namely, byproduct mutualism (Brown, 1983), however, may be as important, if not more important, than these three other routes.

After outlining how the prisoner's dilemma works and where it may be applicable, Brown (1983) introduces the concept of byproduct mutualism:

> In nature it is likely that many payoff matrices for potential cooperators depart from the requirement of the Prisoner's Dilemma. I predict that in many cases of mutualism, CC > DC will be found to prevail rather than DC > CC as required by the prisoner's dilemma. A few examples illustrate the point.
>
> In byproduct mutualism, each animal must perform a necessary minimum itself that may benefit another individual as a byproduct. These are typically behaviors that a solitary individual must do regardless of the presence of others, such as hunting for food. In many species these activities are more profitable in groups than alone, so that CC > CD > DC ≅ DD. In other words, consistent defection (meaning depending completely on others) is impossible or foolhardy. (p. 30)

Following Brown (1983), Connor (1986) introduced the concept of pseudo-reciprocity, which he describes as "investing in byproduct mutualism." Although it may prove true that pseudo-reciprocity is more likely to be found in the context of *inter-specific* interactions (an issue beyond the scope of this book), it is nonetheless an interesting idea that applies as well to intraspecific interactions (see Connor, 1986, 1995a) and is well worth exploring. Connor (1986, 1995a) argues that many actions that appear to represent reciprocity do not, and proposes:

> a solution to this problem that involves a kind of interaction that resembles reciprocity, but which entails no probability of cheating, and therefore is much more likely to occur in a variety of species. . . . Consider the following hypothetical scenario: at some cost to itself, a bird whose nest is in a hot, sunlit area spends some time and effort "fertilizing" a nearby bush to cause it to grow taller and shade the bird's nest. However unlikely this scenario, it clearly illustrates the pseudo-reciprocity concept: the bird has performed a beneficent act for the bush, but the bush has not performed a beneficent act for the bird in return. Rather, the bird's return benefit is an incidental effect or byproduct of a self-promoting act (growth) on the part of the bush. (1986, p. 1562)

Byproduct mutualism is a kind of paradox in the world of cooperation. This category might be thought of as the simplest type of cooperation in that no kinship need be involved, nor are the cognitive mechanisms that require score-keeping or the population structure required for group-selected cooperation necessary for byproduct mutualism to evolve. As such, byproduct mutualism is "simple" in the sense of what is needed for cooperation to evolve, and this in turn might make it the most common category of cooperation, when all is said and done. Nonetheless, some might argue that if you are interested in cooperation in animals, don't even bother documenting how common byproduct mutualism is, because it is not cooperation to begin with. Proponents of such a view would hold that since cheaters do worse than cooperators under this category of cooperation, there is no temptation to cheat and so cooperation is a misnomer. I believe that such a view of cooperation is overly restrictive and depends too heavily on what might be thought of as a "prisoner's dilemma view of cooperation." Because byproduct mutualism requires some sort of coordinated action on the part of the players involved and because high payoffs cannot be obtained without the presence of other cooperating individuals, I consider it to be a respectable category of cooperation. To establish this point, suffice it to say that if we took the logic of byproduct mutualism and applied it to genes within individuals, rather than individuals within dyads, most of what is considered "classic" individual-level organization would qualify as byproduct mutualism!

The preceding discussion suggests that we might consider changing the name of this category of cooperation from byproduct mutualism to "no-cost" cooperation for two reasons. First, the phrase byproduct mutualism in and of itself seems to trivialize this category of cooperation, and, second, the latter title better reflects the evolutionary processes underlying this category of cooperation. Let us address each of these concerns in turn, starting with the trivialization issue. It could be argued that the term byproduct mutualism better reflects the psychological aspects of this category of cooperation than the evolutionary aspects. Imagine two lions and a large gazelle on the plains of Africa (we return to this example in its true form in Chapter 5). Further imagine that the only way to capture and kill the gazelle is if both lions hunt together. Now consider this problem from the viewpoint of lion 1 (the same will hold true for lion 2). The lion has two choices: hunt (cooperate) or don't hunt (don't cooperate). If it hunts, it shares in the gazelle, and if it doesn't, it remains as hungry as it was. Thus lion 1 pays an immediate cost for not hunting and hence decides to hunt. From this perspective, lion 1 makes a decision based on how to get the greatest benefit for the least cost and, in so doing, opts to cooperate. The fact that lion 2 obtains a benefit as well is incidental—it is a byproduct of the actions of lion 1's attempt to get the most it can, and its decision was made without any reference to lion 2 (see Nunney, 1985, for more on this sort of reasoning). From this perspective, the term byproduct mutualism nicely captures the "psychological calculus" of the situation. But we must be wary of mixing evolutionary and psychological explanations.

No doubt, the last two sentences of the above paragraph will make some readers believe it is me who has lost track of the difference between evolutionary and psychological questions. After all, couldn't one recast the argument and say that *natural selection* favors lion 1 to cooperate, and that the fact that others benefit is a byproduct of this evolutionary process? Absolutely! My argument is simply that since psychological calculus is so easy to use, so compelling at first glance, and is so often invoked when describing byproduct mutualism and (even other types of cooperation; Nunney, 1985), we must be very vigilant when using a term such as this.

Is there a term that might better capture the essence of what has been called byproduct mutualism? One possibility is no-cost cooperation (I thank David Sloan Wilson for suggesting this term). Let us return for a moment to the lion-hunting example outlined above. The most extreme case that might qualify as byproduct mutualism, but which we might temporarily refer to as no-cost cooperation, is shown in Figure 2.2. In this case, unless a pair of lions cooperate when hunting, their payoff is always zero. It seems reasonable to refer to this as no-cost cooperation. The trouble is that this is a slightly bizarre payoff matrix in that, if defection is the primitive state, then cooperation does not spread except via drift. Let us consider a slightly different matrix, but one that still qualifies as no-cost cooperation. In Figure 2.3, CC still has the highest payoff entry, but now individuals in CD groups obtain a better payoff than individuals in DD groups, and C can invade from mutation frequency.

The term *no-cost cooperation* has its benefits in that it is not laden with the problems associated with the term *byproduct mutualism* and perhaps is closer to an operational definition. Whether one adopts this terminology is not as important as the issues that terms like *byproduct mutualism* call to the surface. In fact, even after all of the above, I will continue to call this category *byproduct mutualism* for

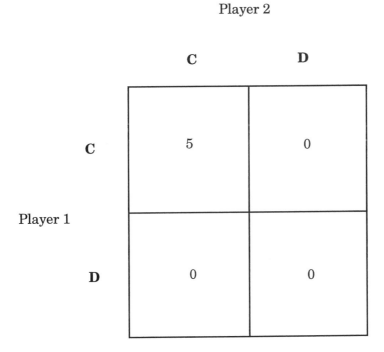

Fig. 2.2. One payoff matrix for no-cost cooperation. Here C can evolve only after drifting into a population of D (i.e., DD = CD = DC).

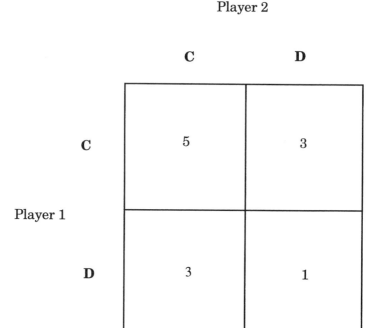

Player 2

Player 1

	C	D
C	5	3
D	3	1

Fig. 2.3. Another payoff matrix for no-cost cooperation. Now C can invade a population of D when at mutation frequency.

one overriding reason—history. This term is already in the literature, and while it has all the problems I have outlined, it seems unwarranted to introduce yet another term to describe a behavior that already has too many labels.

2.6 The cooperator's dilemma game

Despite the fact that all four categories outlined above can lead to the evolution of cooperative behavior, until recently no attempt had been made to link all four categories under a single theoretical umbrella. As such, my colleague, Michael Mesterton-Gibbons, and I (Dugatkin et al., 1992; Mesterton-Gibbons and Dugatkin, 1992; Dugatkin and Mesterton-Gibbons, 1994) have tried to develop a framework for studying all four paths to cooperation using a single game—a game we call the cooperator's dilemma.

To introduce the cooperator's dilemma, I begin by noting that actions to be distinguished as cooperative or noncooperative must be part of an *inter*action among two or more individuals. The simplest possibility is a two-person interaction, and the simplest model of it is the symmetric two-strategy game shown in Figure 2.4. In this figure $\rho > \pi$, $2\rho > \sigma + \tau$, C and D denote the strategies, and the matrix yields the payoff in terms of fitness to player 1. Let UV denote the strategy combination in which the player 1 selects U and its opponent selects V. Then the payoffs associated with the four possible strategy combinations CC, CD, DC, and DD are, respectively, ρ, σ, τ, and π to player 1, and ρ, τ, σ, and π to player 2; and so the combined payoffs are, again respectively, 2ρ, $\sigma + \pi$, $\pi + \sigma$, and 2π. Since CC yields the best combined payoff, which can be achieved only if both individuals select C, we will call C a cooperative strategy, and D a strategy that codes for noncooperation, or defection. This game is referred to as the cooperator's dilemma.

If it is also true that $\tau > \rho$ and $\pi > \sigma$, then the cooperator's dilemma becomes the prisoner's dilemma, or PD, the most celebrated model of the difficulties of achieving cooperation among the members of a group (see Chapter 1). It will often be convenient to use alternative notations for payoffs when the cooperator's dilemma becomes the prisoner's dilemma. When this occurs, we will replace ρ, σ, τ, and π by their respective English analogs, namely, R, S, T, and P (as in Fig. 1.1). Prisoner's dilemma calculations are greatly simplified if payoffs satisfy

$$S + T = P + R \qquad (2.11)$$

a condition that holds, for example, in Caraco and Brown (1986), in Newman and Caraco (1989), and in section 2 of Mesterton-Gibbons (1991). Suppose that player 1 cooperates. Then it obtains R if player 2 cooperates, but only S if it

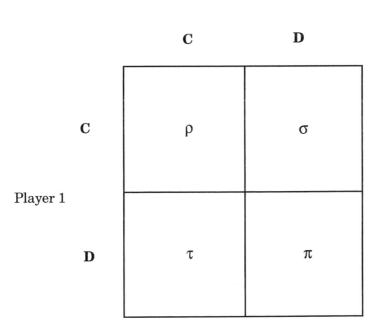

Fig. 2.4. The payoff matrix for the cooperator's dilemma (after Mesterton-Gibbons and Dugatkin, 1992).

defects; therefore the benefit to player 1 of player 2's cooperation is $R - S$. Suppose, on the other hand, that player 1 defects. Then it obtains T if its opponent cooperates, but only P if its opponent defects; therefore the benefit to player 1 of player 2's cooperation is $T - P$. But $R - S = T - P$ when 2.11 is satisfied, and so we can define

$$b = R - S = T - P \qquad (2.12)$$

to be the benefit that player 1 receives as a result of its opponent's cooperation—regardless of player 1's strategy. Similarly, regardless of player 2's strategy, we can define

$$c = T - R = P - S \qquad (2.13)$$

to be the cost to player 1 of its own cooperation. Note that the benefit b and cost c of cooperation satisfy $b > c > 0$. When (2.11) is satisfied, the prisoner's dilemma is said to be *decomposable*. For further elaboration of this and all other game-theoretic concepts used in this chapter, see Mesterton-Gibbons (1992a).

Let us now import the four categories of cooperation outlined earlier (reciprocity, group selection, kinship, and byproduct mutualism) into the cooperator's dilemma and introduce some terminology as well. Within each of the four categories, there is a principal effect whose removal would select for noncooperative behavior. Following Mesterton-Gibbons and Dugatkin (1992), I will refer to this effect as the *mechanism* for cooperation, and if I identify a further effect whose removal would also select for noncooperative behavior, then I will refer to it as a *prerequisite* for cooperation.

2.6.1 Group-selected behavior

In group-selected behavior, the scope of an interaction is broadened—spatially or otherwise (see Wilson, 1980, for discussion)—to include the indirect effects on an individual's fitness of the productivity of its local population, or trait group. Spatial heterogeneity allows natural selection to act not only on individuals relative to others in their local population, but also on local populations relative to others in the global population, or deme. In such models, even variation caused by the random formation of groups can be enough to ensure that cheating locally is penalized globally. When variation in the frequency of cooperators between groups is greater than random, the group-level penalties for cheating increase. As such, although the incentive to cheat still exists within trait groups, it fails to occur at the level of the deme.

The mechanism for cooperation in group-selected behavior is *deme structure,* that is, variance between trait groups in the frequency of the cooperative strategy C. To see this, consider a decomposable prisoner's dilemma in the simplest possible example of a structured deme, that is, an infinitely large population with two trait groups of equal (and infinite) magnitude. Let x_1 be the frequency of C-strategists in trait group 1, and y_1 be the frequency of D-strategists, so that $x_1 + y_1 = 1$; similarly, let x_2 be the frequency of C-strategists in trait group 2, and $y_2 = 1 - x_2$ the frequency of D-strategists. Let player 1 be randomly assigned to one of these trait groups, in which it has a single prisoner's-dilemma interaction with a randomly chosen partner. If player 1 is a C-strategist, then it is placed into group 1 with probability $x_1/(x_1 + x_2)$ and into group 2 with probability $x_2/(x_1 + x_2)$. If player 1 is a D-strategist, then it is assigned to group 1 with probability $y_1/(y_1 + y_2)$ and to group 2 with probability $y_2/(y_1 + y_2)$; again, the sum of these numbers is 1. The probability that the protagonist will interact with a C-strategist, given that it has been assigned to trait group 1, is the frequency of C within trait group 1 (because the trait groups are infinitely large), namely, x_1. The probability that player 1 will interact with a D-strategist, if it is placed in trait group 1, is the frequency of D within trait group 1, namely, y_1. Similarly, conditional upon being assigned to trait group 2, the probabilities that the protagonist will interact with a C-strategist or a D-strategist are x_2 or y_2, respectively.

Now, the probability that a randomly assigned C-strategist interacts with a D-strategist—which is denoted by p_{CD}—equals [probability it interacts with a D-strategist if assigned to trait group 1] x [probability it is assigned to trait group 1] + [probability it interacts with a D-strategist if assigned to trait group 2] x [probability it is assigned to trait group 2], which equals $y_1 \cdot x_1/(x_2 + x_2) + y_2 \cdot x_2/(x_1 + x_2)$. Similarly, the probability that a randomly assigned C-strategist interacts with a C-strategist is $p_{CC} = x_1 \cdot x_1/(x_1 + x_2) + x_2 \cdot x_2/(x_1 + x_2)$; the probability that a randomly assigned D-strategist interacts with a C-strategist is $p_{DC} = x_1 \cdot y_1/(y_1 + y_2) + x_2 \cdot y_2/(y_1 + y_2)$; and the probability that a randomly assigned D-strategist interacts with a D-strategist is $p_{DD} = y_1 \cdot y_1/(y_1 + y_2) + y_2 \cdot y_2/(y_1 + y_2)$. Now, given that $y_1 = 1 - x_1$ and $y_2 = 1 - x_2$, it can be shown that:

$$p_{CC} = \frac{x_1^2 + x_2^2}{x_1 + x_2} \qquad p_{CD} = \frac{(1 - x_1)x_1 + (1 - x_2)x_2}{x_1 + x_2}$$

$$p_{DC} = \frac{x_1(1 - x_1) + x_2(1 - x_2)}{2 - x_1 - x_2} \qquad p_{DD} = \frac{(1 - x_1)^2 + (1 - x_2)^2}{2 - x_1 - x_2}.$$

Let the expected payoff to the protagonist be f_C if it is a C-strategist or f_D if it is a D-strategist. The payoff to a C-strategist is R if its opponent is a C-strategist or S if its opponent is a D-strategist; and so its expected payoff is R times the probability that it interacts with a C-strategist plus S times the probability that it interacts with a D-strategist, or

$$f_C = Rp_{CC} + Sp_{CD} \qquad (2.14)$$

Similarly, the expected payoff to a D-strategist is T times the probability that it interacts with a C-strategist plus P times the probability that it interacts with a D-strategist, or

$$f_D = Tp_{DC} + Pp_{DD} \qquad (2.15)$$

Cooperative behavior is selected for if $f_C > f_D$. After routine algebraic manipulations, this condition reduces to:

$$b(x_1 - x_2)^2 + c(x_1 + x_2)(x_1 + x_2 - 2) > 0 \qquad (2.16)$$

where b is the benefit, and c is the cost, of cooperation; both are defined by (2.12) and (2.13). It must also be true that:

$$0 \leq x_1 \leq 1, \quad 0 \leq x_2 \leq 1 \qquad (2.17)$$

The region defined by these inequalities—the region in which cooperative behavior is selected for—is shown in Figure 2.5. Correspondingly, noncooperative behavior is selected for if $f_D > f_C$. The greater the amount by which b exceeds c, that is, the greater the amount by which the benefit of cooperation exceeds the cost, the larger the cooperative region in Figure 2.5. The line where $x_1 = x_2$ always lies in the region in which cooperation is selected

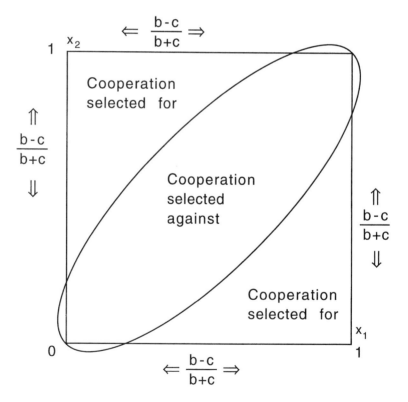

Fig. 2.5. Group-selected cooperation in a structured deme. x_k is the frequency of cooperators in trait group k (after Mesterton-Gibbons and Dugatkin, 1992).

against. Thus C is always selected against in a homogeneous population ($x_1 = x_2$). But C is selected for in a heterogeneous population if the variance between trait groups is sufficiently large, that is, if $|x_1 - x_2|$ is sufficiently large.

A little algebraic manipulation shows that $p_{CC} > p_{DC}$ and $p_{DD} > p_{CD}$ whenever $x_1 \neq x_2$—that is, whenever the variance, $(x_1 - x_2)^2/2$, is positive. In other words, this is the case in which each of the two strategies is more likely to interact with itself than with its competitor. Thus deme structure segregates cooperators and noncooperators. In this model, the variance is caused by chance, and is therefore passive, but the necessary variance can also be generated actively, by cooperators associating with cooperators or expelling defectors from their trait group. Note that if the productivity of a trait group is proportional to the frequency of cooperators, then $x_1 \neq x_2$ also implies *differential trait-group productivity*, which is a general prerequisite for group-selected cooperative behavior.

2.6.2   Reciprocal altruism

In reciprocal altruism, the history of an interaction between two individuals is extended forward in time to account for future encounters. The simplest such model is the iterated prisoner's dilemma, or IPD, in which a PD interaction repeats itself between fixed opponents an unknown number of times, say M, where M is a random variable. In this iterated game, player 1's strategy is an infinite vector $\mathbf{u} = (u_1, u_2, u_3, \ldots)$, where u_k (taking values C or D) is the PD strategy used on interaction k; and actions can be conditional upon the behavior of the opponent, as in TFT, in which a protagonist cooperates during its first encounter, but thereafter does what its opponent did during the previous encounter. Thus,

$$\text{TFT} = (C, v_1, v_2, \ldots) \tag{2.18}$$

where v_k is player 2's strategy on interaction k.

The mechanism for cooperation in reciprocal altruism is "score-keeping" (Brown, 1983): the protagonist conditions its behavior during later encounters on its opponent's behavior during earlier encounters—as, for example, in TFT. Prerequisites for this mechanism to work are: (1) the expected number of interactions between individuals, say μ, must be sufficiently large, and (2) for nonsessile organisms, individuals must have sufficiently well-developed neural apparatis to recognize associates and remember their actions on previous encounters.

To see how score-keeping works, we consider the IPD in which TFT and ALLD are the only strategies. To calculate the payoff matrix, observe that two TFT-strategists will both select C at each interaction, thus each will obtain PD payoff R a total of M times, for an IPD payoff MR. Because M is a random variable, however, this IPD payoff is also a random variable, and so we replace it by its expected value $= \mu R$, where $\mu = E(M)$ is the expected value of M. Similarly, the total payoff to each of two ALLD-strategists is MP, with expected value $= \mu P$. If the protagonist is a TFT-strategist and its opponent an ALLD-strategist, then the protagonist will select C on the first interaction and D thereafter; whereas its opponent will always select D. Thus, the protagonist's total payoff is $S + (M - 1)P$, with an expected value of $S + (\mu - 1)P$; whereas the opponent's expected total payoff is equal to $T + (\mu - 1)P$. Figure 2.6 shows the payoff matrix for such a game.

Let us assume for simplicity's sake that $S + T > 2P$, which is true if the PD is decomposable. Then this payoff matrix satisfies $2\rho > \sigma + \tau, \rho > \pi$, so the game is still a cooperator's dilemma; moreover, one of the inequalities of the prisoner's dilemma still holds, namely, $P > S$. But it is no longer necessarily true that T exceeds R. Rather, if

$$\mu > \frac{T - P}{R - P} \qquad (2.19)$$

then $\rho > \tau$, even though R < T; that is, although the incentive to cheat still exists at the level of the present time, it now fails to exist at the level of the indefinite future. The noncooperative strategy (ALLD) is no longer a dominant strategy, yet neither is it a cooperative one (TFT); rather, each is a best reply to itself. In other words, both TFT and ALLD are evolutionarily stable strate-

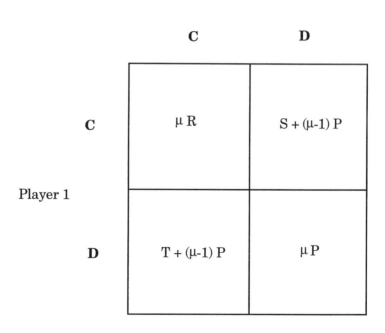

Fig. 2.6. The payoff matrix for a game of TFT versus ALLD (see text for details).

gies (or ESSs) in the sense of Maynard Smith (1982). Thus score-keeping can maintain cooperation under natural selection, but cannot initiate it.

2.6.3   Byproduct mutualism

A prerequisite for byproduct mutualism (Brown, 1983; also see West-Eberhard, 1975, and Rothstein and Pirotti, 1987, for a discussion of similar concepts) is what we will call the *boomerang factor*—any uncertainty that increases the probability that a defector will be the victim of its own cheating. The typical interaction is still described by the cooperator's dilemma, but now $\rho - \tau$ and $\sigma - \pi$ are increasing functions of an exogenous parameter δ, which measures the adversity of the environment (see Fig. 2.7). There are two critical values of δ, labeled δ_1 and $\delta_2(\geqslant \delta_1)$. If $\delta < \delta_1$, both $\rho - \tau$ and $\sigma - \pi$ are negative and the game is a prisoner's dilemma. If $\delta_1 < \delta < \delta_2$, however, then $\rho - \tau > 0 > \sigma - \pi$; both C and D are ESSs. Finally, if $\delta > \delta_2$, then $\rho - \tau > \sigma - \pi > 0$; C is a dominant strategy (and the sole ESS). Thus selection is for defection when $\delta < \delta_1$, but for cooperation when $\delta > \delta_2$. The "*common enemy*" of a sufficiently adverse environment provides a mechanism for cooperation (Mesterton-Gibbons, 1991).

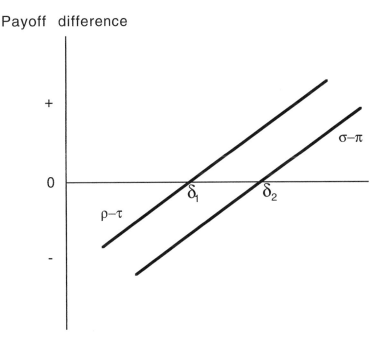

Payoff   difference

Fig. 2.7.   The common enemy of a sufficiently adverse environment. The increase in the payoff differences $\rho - \tau$ and $\sigma - \pi$ with δ, which measures environmental adversity, is represented schematically. (The increase need not be linear and the curves need not be parallel) (after Dugatkin and Mesterton-Gibbons, 1992).

A rapid transition from D to C as δ increases is also a characteristic feature of three recent models which use this common-enemy mechanism; $\delta_1 - \delta_2$ is zero in the first two cases (so that the curves in Fig. 2.7 coincide) and small (but positive) in the third. Caraco and Brown (1986) describe food sharing (versus D = not sharing) when food is abundant and applied the model to the Mexican jay (*Aphelocoma ultramarina;* a separate nesting, plural breeding species). The boomerang factor here is the proximity of fledglings, with the consequent danger that a predator's attention may be drawn to the protagonist's young if the opponent's young beg too loudly for food and δ is the risk of predation on fledglings. In a similar vein, Newman and Caraco (1989) describe calling (versus D = not calling) to alert conspecifics when food is abundant and a patch has been discovered. The boomerang factor is that predation risk for a solitary forager exceeds the risk for a social forager; δ is the risk of predation. Mesterton-Gibbons (1992a) describes superparasitism (versus C = not superparasitizing) when oviposition sites are abundant for foraging insects. The boomerang factor is unrecognition, with the consequent danger that an insect may superparasitize its own site; and δ is the ratio of a solitary egg's survival probability to that of a paired egg.

Despite the appearance of the IPD in the title of his paper, Lima's 1989) example of cooperation in an "antipredatory game" is essentially a case of byproduct mutualism. His boomerang factor is the individual's risk of losing its partner through cheating (even, as it were, to defect against), and the common enemy is the high risk of predation while in solitude. Another possible example of a common enemy—"floaters" in search of a territory—appears in Getty's (1987) discussion of defensive coalitions among territorial neighbors. Here the boomerang factor is the cost in fitness of renegotiating boundaries with new, unfamiliar neighbors (this example may also contain elements of reciprocity as well).

2.6.4   Kinship

The cooperator's dilemma can be accommodated to study kin-selected behavior, by calculating what Hines and Maynard Smith (1979) refer to as the "Grafen ESS." In kin-selected behavior, the mechanism for cooperation is common ancestry (i.e., genes that are identical by descent: Hamilton, 1964a, b), with the prerequisite for cooperation being a sufficiently high degree of relatedness between players.

Early attempts to incorporate kinship into game theory (e.g., Mirmirani and Oster, 1978) argued that standard entries into a payoff matrix needed to be modified such that expected entry in a payoff matrix, $E(X, X)$, becomes $E(X, X) + rE(Y, X)$, where r represents the coefficient of relatedness between players. Grafen (1979), however, points out that this approach contains a fallacy, in that although the payoff matrix itself is modified correctly, this approach does not take into account the fact that in games between relatives a player using strategy X has a greater than random chance of meeting another X strategist. Hines and Maynard Smith (1979) then calculate the "Grafen ESS" for games between kin using Grafen's "personal fitness" approach.

Let X and Y represent the two strategies used in a game between relatives, and assume that all but some proportion ϵ of the population uses strategy X. Following Grafen (1979), let r now equal the probability of meeting (haploid, asexual) kin, and let $(1 - r)$ be the probability of encountering a randomly chosen player in the population. Then Hines and Maynard Smith (1979) show that the mean expected payoff to these strtegies are

$$W(X) = rE(X, X) + (1 - r) [(1 - \epsilon) E(X, X) + \epsilon E(X, Y)] \quad (2.20)$$

$$W(Y) = rE(Y, Y) + (1 - r) [(1 - \epsilon) E(Y, X) + \epsilon E(Y, Y)] \quad (2.21)$$

and that X is an ESS if

$$E(X, X) + (1 - r)\epsilon[E(X,Y) + r(E(X, X)] >$$
$$(1 - r)E(Y, X) + rE(Y, Y) + (1 - r)\epsilon [E(Y, Y) - E(Y, X)] \quad (2.22)$$

(which for small ϵ requires $E(X, X) \geqslant (1 - r)(E(Y, X) + rE(Q, Q))$ and

$$E(X, Y) - E(X, X) - E(Y, Y) + E(Y, X) > 0 \quad (2.23)$$

Hines and Maynard Smith (1979) then use these equations to calculate the Grafen ESS for the hawk-dove game. Calculating the ESS for the cooperator's dilemma begins by modifying the payoff matrix as follows:

	C	D
C	$(1 + r)\rho$	$\sigma + r\tau$
D	$\tau + r\sigma$	$(1 + r)\pi$

Necessary conditions for C to be an ESS are then:

$$(1 + r)\rho > \tau + r\sigma \quad (2.24)$$

and

$$(1 + r)\rho > \tau + r\sigma + r(\rho + \pi - \tau - \sigma) \quad (2.25)$$

Necessary conditions for D to be an ESS are then:

$$(1 + r)\pi > \sigma + r\tau \quad (2.26)$$

and

$$(1 + r)\pi > \sigma + r\tau + r(\rho + \pi - \tau - \sigma) \quad (2.27)$$

Furthermore, Hines and Maynard Smith (1979) show that if $\rho + \pi > \sigma + \tau$, then no mixed strategy is a Grafen ESS, and at least one of C or D is an ESS.

If $\rho + \pi > \sigma + \tau$ then C is an ESS if $(1 + r)\rho > \tau + r\sigma$, D is an ESS if $(1 + r)\pi > \sigma + r\tau$. Otherwise, play D with probability

$$\frac{(\sigma - \pi) + \rho(\tau - \pi)}{(1 + r)(\sigma + \tau - \rho - \pi)} \tag{2.28}$$

is an ESS.

A caveat is in order here: although it can be shown that kin selection is a subset of group selection (Wilson, 1980, 1983; Wilson and Sober, 1994; Dugatkin and Reeve, 1993), for the purposes of delineating mechanisms of cooperation and categorizing behaviors by these mechanisms, I think it justified to separate kin-selected behavior from group-selected behavior. That is, even though from a gene-counting perspective (such as that of Dawkins, 1976, 1989) kin and nonkin group selection are equivalent, from a behavioral perspective the proximate factors associated with kin and nonkin group selection are likely to be quite different. In addition, this separation proves to be useful because the coefficient of relatedness r has been calculated for populations of many species, while this is not the case for the equivalent association parameter in nonkin group selection. So, while perhaps I should have technically called the group-selected category of cooperation "nonkin group selected" cooperation, I cringe at the idea of yet more terminology, and hence keep the original title.

Finally, it is only fair to note that the choice of terminology in distinguishing mechanisms from prerequisites is a bit subjective. Where there are two or more features whose removal would select for noncooperative behavior, the intent has been to label as mechanism the most active feature—that which most evokes the power of enforcement—and hence label as prerequisites the more passive ones.

2.7   Onward and upward

In the next five chapters, we will examine cooperation in a wide array of taxa (insects, fish, birds, and mammals) within the framework described above. Although the four categories outlined so far are quite broad and despite the fact that it will be duly noted that many examples do not fall neatly into any of the four categories (but rather straddle the fence between two or more), I will no doubt be accused of trying to pigeonhole every case of cooperation into the cooperator's dilemma framework. This is not an unreasonable accusation—such is the price that must always be paid when developing broad conceptual frameworks.

3

Cooperation in fishes

For Aristotle in *Historia Animalium,* the motivation of fish ranged from enjoyment of tasting and eating to madness from pain in pregnancy, and for Francis Day, summarizing for the 1878 Proceedings of the London Zoological Society in wake of Darwin's epochal *The Expression of the Emotions in Man and Animals,* fish were variously moved by disgrace, terror, affection, anger and grief. (Colgan, 1986, p. 23)

3.1   Introduction

Every so often behavioral ecologists need to remind themselves that there are over 22,000 species of teleost fishes alone (Pitcher, 1986)—more than all other vertebrates combined! Many of these species take well to the laboratory, where they reproduce quickly and proficiently. It should therefore come as no surprise that some of the best-controlled studies in optimal foraging theory, sexual selection, sexual allocation, parental care, and, yes, cooperative behavior have been undertaken using fishes (see Pitcher, 1986; Godin, 1996). That being said, it may seem strange that fish are so commonly used for research on cooperation. After all, fishes are "lower" vertebrates and were thought by many, therefore, to have a relatively simple behavioral repertoire. Evidence, however, shows this view to be manifestly false (Dugatkin and Wilson, 1993; Dugatkin and Sih, 1995) and indicates that some teleost species contain a rich repertoire of complex social behaviors, many of which appear to be very labile. In this chapter, I will examine cooperation in teleost fishes in the context of egg trading, helpers-at-the nest, "cleaning" behavior, cannibalism, foraging, alarm substances, mobbing behavior, and predator inspection and try, whenever possible, to place these studies within the framework of the cooperator's dilemma.

3.2   Egg trading and reciprocity in simultaneously hermaphroditic fishes

The probability of finding an example of cooperation via reciprocity would seem to be at its greatest when cooperators produce physical goods that are exchanged. That is, although any behavior can in principle be reciprocated, if the action itself involved the production and exchange of some physical, measurable good, reciprocity would be at least that much easier to quantify. In some simultaneously hermaphroditic fishes (and worms), just such an exchange takes place.

Among the vertebrates, simultaneous hermaphroditism—the possession by a single individual of both eggs and sperm at the same time—is relatively rare and occurs only in fishes. Within the fishes, it is most often found in deepwater species with low population densities (Smith, 1975), as predicted by sex allocation theory (Tomlinson, 1966; Charnov, 1982), but has been best documented in the seabasses, a shallow-dwelling taxa (Clark, 1959; Robins and Starck, 1961; Smith, 1965; Barlow, 1975; Fischer, 1980, 1984). In the seabasses, individuals spawn daily:

> Individuals form pairs late in the afternoon. Before spawning, two mates alternate courtship displays, and the last fish to display releases eggs, while its partner releases sperm (fertilization is external and eggs are planktonic). Each fish releases only part of its clutch during a given spawning act, and partners regularly alternate release of eggs. (Fischer, 1988, p. 120)

Clutches of eggs in at least eight serranine (seabass) species are divided in parcels, and partners in such cases exchange roles (producing egg parcels and producing sperm) about 80% of the time (Fischer, 1984, 1986; see Warner, 1984; Peterson, 1987, 1990, 1991; Leonard, 1993, for more on gender allocation in some serranids). Individuals in one of the better studied species, *Hypoplectrus nigricans,* fertilize, on average, about as many eggs as they parcel out (Fischer 1980; but see Peterson, 1995, for evidence that relative size affects egg-trading frequency in the tobaccofish, *Serranus tabacarius*). Fischer (1988) refers to such egg swapping as "delayed reciprocity." If this is a real case of reciprocity, however, and eggs are more expensive to produce than sperm, then why not cheat, that is, produce sperm, but fail to parcel out eggs? (Leonard, 1990.) The answer might lie in the notion that simultaneous hermaphrodites are trapped in an iterated prisoner's dilemma, and are employing something akin to TFT when paired up with a partner. Fischer (1988) suggests this possibility and outlines what one might look for, if it were in fact the case:

> To determine if egg trading is an evolutionarily stable form of TfT requires four steps. First, mating behavior in the serranines must satisfy the conditions of the iterated Prisoner's Dilemma. Second, animals must exhibit some form of TfT. Third, the probability of an additional interaction in a given game must be large enough to make this form of TfT evolutionarily stable. Fourth, the TfT behavior cannot be accounted for by alternative mechanisms. (p. 124)

Does egg trading meet these demands? Let us begin to answer this question by defining a game as that sequence of mating interactions that occurs during a session of spawning (Fischer, 1988). An "iteration," or play of the game, is then the release of parcels by one player and the subsequent release of, or failure to release, eggs by its partner. Following Fischer (1988), let cooperation be equated with parceling eggs on a player's appointed turn and defection equal the converse. If b = the benefit of fertilizing eggs, c = the cost of producing eggs, and p = the probability that a defector *does* give up a parcel on its appointed turn, then the payoff matrix for such a game is shown in Figure 3.1.

This matrix would meet the conditions of the prisoner's dilemma if $p < 1$ and B > C > 0. The first of these conditions guarantees $T > R$, $P > S$ and $2R > T + S$ and the second condition assures that $R > P$. TFT becomes a stable solution to this game whenever the probability of future interaction, w, is greater than c/b.

Are these inequalities met by egg swappers? By definition $p < 1$, and although the cost of producing eggs has not been measured, Fischer (1988) argues that evidence suggests that the cost of producing eggs (c) is relatively small compared with the benefit of fertilizing them (b). If this proves to be the case, then the payoff matrix for egg-swapping qualifies as a prisoner's dilemma, and w (the probability of future interaction with the same the partner) need only be small for TFT to be one solution to the game.

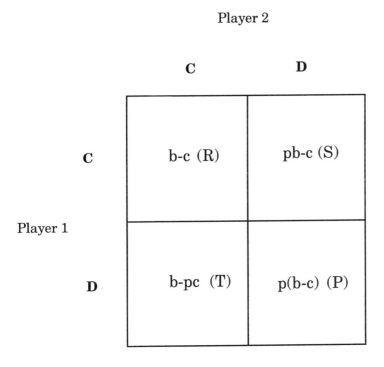

Player 2

	C	**D**
C	b-c (R)	pb-c (S)
D	b-pc (T)	p(b-c) (P)

Player 1

Fig. 3.1. The payoff matrix associated with egg parceling in simultaneous hermaphrodites. b = the benefit of fertilizing eggs, c = the cost of producing eggs, and p = the probability that a defector *does* give up a parcel on its appointed turn. T, R, P, and S are the equivalent entries for the prisoner's dilemma game (after Fischer, 1988).

Despite the lack of systematic studies, some evidence exists that simultaneous hermaphrodites that engage in egg swapping use something akin to TFT. For example, Fischer (1980; and Fischer, unpublished data, cited in Fischer, 1988) provides some data that suggest black hamlets (*Hypoplectrus nigricans*) and chalk bass (*Serrannus tortugarum*) retaliate against cheating. In these species, mates waited significantly longer to parcel out eggs to a partner who had failed to reciprocate in prior interactions. Fischer (1988, p. 125) notes, however, that "it could be that egg-trading is nicer or more forgiving than simple TFT." It seems, in fact, that rather than playing pure TFT, egg swappers use something more akin to Generous TFT (Nowak and Sigmund, 1992), in that individuals seem to reciprocate only about 80% of the time, yet pairs often stay together for long periods of time.

Axelrod and Hamilton (1981) suggest that one way for TFT to invade a population is for interactions to be among kin early in the evolutionary trajectory. This logic, however, fails to explain the initial spread of TFT during egg swapping, as eggs in these species are typically planktonic and drift long distances, making kin selection unlikely. Rather than kin selection, the evolution of reciprocity in egg swappers appears to follow a path similar to that set out in a model developed by Peck and Feldman (1986), in which the payoff matrix does not begin as a prisoner's dilemma, but rather develops into one through time (Fischer, 1988).

Fischer (1988) argues that early on, conditions (such as low density) favored alternating egg production, but not parceling eggs. The game, at this early stage, did not, however, meet the conditions of the prisoner's dilemma. Then, conditions changed and, for example, mate density might have increased to a point at which parceling evolved because it led to high mating success. "Once it was common, the combination of egg-offering, parceling and waiting for the mate to reciprocate satisfied the conditions of TfT" (Fischer, 1988). Fischer further extends this argument and proposes that once TFT is in place, individuals do just as well when sticking with one mate as when switching, thus favoring fidelity in relationships. Although many of the assumptions behind this scenario remain to be tested, it is nonetheless an interesting idea, and one of the few attempts to explain the origination, rather than the maintenance, of TFT in a population.

A remarkably similar form of egg trading and reciprocity occurs in the protandrous, hermaphroditic polycheate worm *Ophryotrocha diadema* (Table 3.1; Sella 1985, 1988, 1991). What makes this case especially fascinating is that as simple a creature as a polycheate worm is able to assess whether its partner is cheating and then respond with an appropriate behavior (Sella, 1988).

Connor (1992) presents an interesting alternative explanation of egg swapping, based on his pseudo-reciprocity model (Connor, 1986; also see Friedman and Hammerstein, 1991, for more on alternative explanations for egg swapping). As opposed to true reciprocal altruism where it always pays an individual to cheat once it has received benefits, no such temptation exists in pseudo-reciprocity:

Table 3.1. The spawning sequence from five pairs of *Ophryotrocha diadema* (a simultaneously hermaphroditic polycheate worm). Two phenotypes were used to allow identification of eggs. Yellow phenotypes lay yellow (Y) eggs, and yy lay white eggs (W). Notice the remarkable degree to which egg-laying roles are alternated (after Sella, 1985).

		Day in which spawing occurred													
Pair number	Partner	1	2	3	4	5	6	7	8	9	10	11	12	13	14
5	YY	Y				Y				Y				Y	
	yy			W		W					W		W		
11	YY	Y				Y				Y			Y	Y	
	yy			W				W			W				W
8	YY		Y			Y		Y				Y			
	yy	W			W			W				W			W
9	YY	Y		Y		Y			Y			Y			
	yy			W	W						W		W		
12	YY		Y			Y					Y		Y		
	yy	W			W					W		W			

In the pseudo-reciprocity paradigm, an individual *A* performs a beneficent act for an individual *B* in order to increase the probability of receiving incidental benefits from *B*. Because the return benefits to *A* derive from behaviors *B* performs to benefit *B*, there is no cheating in pseudo-reciprocity. (Connor, 1992, p. 523)

Connor (1992) presents a simple model of pseudo-reciprocity in which an individual decides on each of *N* moves whether to give a "parcel," 1/N, of the total benefits it is able to produce to its partner. Following his nomenclature, let:

B_s = the benefit of staying in the interaction
C_s = the costs of staying in the interaction
B_l = the benefit of leaving
C_l = the costs of leaving

Connor (1992) proposes that simultaneous hermaphrodites parcel out eggs in an attempt to make $R > T$ ($B_s - C_s > B_l - C_l$) and thus engage in pseudo-reciprocity, rather than true reciprocity. Even in pseudo-reciprocity, however, the temptation to cheat on the last move (if it is known to be such) still exists. In simultaneous hermaphroditic egg swappers such as *S. tortugarum* and *H. nigricans,* however, eggs must be used the same day they are produced, and this may make parceling eggs worthwhile even on the last move of a given day.

Since differentiating TFT and pseudo-reciprocity relies on knowing B_s, B_l, C_s, and C_l, until these parameters are quantified, no resolution on this question can be reached. A comparison of *S. tortugarum* and *H. nigricans,* however, suggests that the conditions for pseudo-reciprocity may hold true in these species. When the temptation to cheat (T) is high, individuals need to boost up the reward for mutual cooperation (R), which can be accomplished by decreasing the size of parcels (Connor 1992). In support of this prediction, *S. tortugarum*, a species that exists in large aggregations, and hence probably suffers a lower cost for leaving a given partner to find another (i.e., a high T), produces twice as many egg parcels as *H. nigricans,* a species which lives in smaller aggregations (Connor, 1992).

Fischer's experimental and theoretical work and Connor's alternative hypothesis for egg parceling suggest that game theory models of cooperation need to be modified, in the following sense. Such models typically assume that the probability of interaction (w) is not a parameter under the control of players in the game. The potential to divide eggs into different numbers of parcels not only suggests that w is under the control of players in this game, but that the payoff matrix entries themselves are affected by such control. That is, Connor (1992) suggests that the actual value that T or R may take on in the cooperator's dilemma is a function of w, which is at least partly under the control of the individuals involved in these interactions. Future theoretical and empirical work is badly needed to determine the extent to which this is true in other systems, and, if it is, what affect it has on the evolution of cooperation.

3.3   Helpers-at-the-nest

In many species of animals, young adults forgo any chance at reproduction. In some, nonbreeding individuals remain in their natal nests and sometimes help raise siblings. Although such helpers-at-the-nest are usually associated with communally breeding birds and mammals (see Chapter 4–6), this phenomenon has been studied in a few species of fish as well (Taborsky and Limberger, 1981; Taborsky, 1984, 1985, 1987). What makes this behavior particularly interesting in fishes is that cooperation seems to have evolved via a combination of kinship, reciprocity, and byproduct mutualism.

When addressing the question of cooperation and helpers-at-the-nest it is important to separate two related questions: (1) Why do young that are physiologically able to breed remain in their natal nest?, and (2) Why should young who stay at home help? Only when we can answer both questions can we truly understand the evolution of cooperative helpers (Brown, 1987).

Given that helpers-at-the-nest are sexually mature—they can produce sperm or eggs—why should they stay at home rather than attempt to find a member of the opposite sex and mate? While the evidence from communal breeding studies in birds suggests that habitat saturation is often an important factor in a helper's decision to stay home (Brown, 1987), habitat saturation appears to play a smaller role in fish helper systems. In fact, Taborsky (1985) found that when potential helpers in *Lamprologus brichardi* were given a choice between staying on their natal territory or going to an unoccupied territory, they chose to stay home. In *L. brichardi* (and quite possibly in many other fish species with helpers; McKaye and McKaye, 1977) it seems that predation pressure off the natal territory is the driving factor in the decision to stay home. For *L. brichardi* the probability of survival off the natal territory when fish are small is quite low (Taborsky, 1984, 1985), and even though young grow slower when they remain at home and help (Fig. 3.2), the threat of predation is apparently strong enough to force the decision.

In an eloquent series of experiments, Taborsky (1984) examined the costs and benefits of helping in *L. brichardi* by addressing multiple hypotheses of why young stay at the nest, clean eggs, and are actively involved in territory maintenance and defense. One potential cost to helping is the slower growth rates of helpers compared with nonhelpers of similar size. Indeed, Taborsky (1984) found that helpers grew more slowly than family-independent fish of equal size (Fig. 3.2). However, it does not seem to be the energy expenditure involved in helping that causes the decreased growth, as equal-sized fish with their own territories and who were raising their own young grew faster than helpers. Rather, it is the helpers' low rank in the male-female-helper hierarchy that seems to deter growth (Taborsky, 1984).

The possible benefits of helping include rearing closely related kin, the experience of raising a brood, safety, and so on (Taborsky, 1984). After a series of experiments addressing a number of putative benefits of helping, Taborsky (1984, 1985) found that the kinship element was the only measurable benefit of helping. That is, helpers did not have more success raising their own broods in the future, they did not take over parental territories, and cannibalism on the part of helpers, while it did take place, could not explain why helpers helped. The kinship benefit attained by helpers manifested itself in the greater number of eggs being produced by parents when helpers were present (Fig. 3.3; Taborsky, 1984).

Parents may allow their young to remain on the territory because they pose

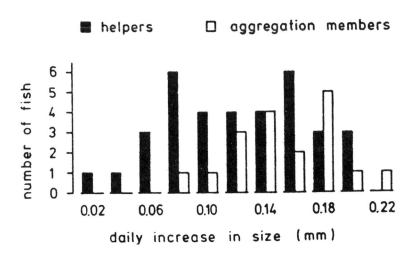

Fig. 3.2. Frequency distribution of helpers and aggregation members of similar size (after Taborsky, 1984).

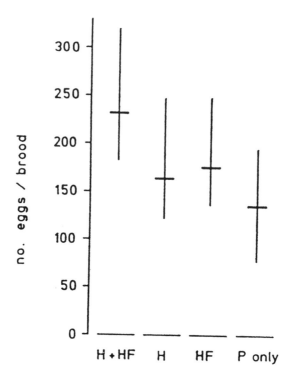

Fig. 3.3. Median clutch size of females with and without helpers. H = helpers, HF = helper fry, P only = Parents only (after Taborsky, 1984).

little in the way of reproductive competition (at least while they are small) and help protect the territory (Taborsky, 1985). When potential reproductive competition from helpers (which increases with the age of the helper) outweighs the benefits and the additional aid in defending territories (and the kinship benefit accrued by letting helpers help), parents expel their helpers (Taborsky, 1984, 1985). This suggests that reciprocity plays a role in parent-helper interactions (but see Soler et al., 1995, for more on similar behaviors that might be better interpreted as manipulation). That is, parents can recognize their own helpers (Hert, 1985) and appear to have a "social contract" (Reeve and Nonacs, 1992) that allows helpers to remain, provided they don't eat eggs. Once helpers start feeding on eggs, females retaliate, attack, and possibly even expel helpers.

Taborsky and his colleague's work on cichlid helpers-at-the-nest is an excellent example of experimentally disentangling different categories of cooperation. As is evidenced in this work, various categories of cooperation may simultaneously be operating on different behavioral decisions, which in sum produce complex behavior such as helping-at-the-nest. In Taborsky's work, we have seen that byproduct mutualism may underlie why helpers stay at home to begin with, kinship may determine why helpers help, and reciprocity might be involved in the decision of how long helpers help.

3.4   Nest raiding in three-spined sticklebacks

Perhaps because of our experiences in human society, we often do not picture "gangs" of females cooperating to overpower males. Yet this is precisely what happens in the context of nest guarding in three-spined sticklebacks (*Gasterosteus aculeatus*). Stickleback parental care is undertaken by males alone (Wooton, 1976), and females appear to form large groups of up to a few hundred that raid and destroy the nests of males guarding conspecific eggs, often eating the eggs contained in such nests (Abdel-malek, 1963; Rohwer, 1978; Kynard, 1979; Snyder, 1984; Whoriskey and FitzGerald, 1985). The possible benefits accrued from nest raiding include: (1) the nutritional value of the eggs eaten (of course, the relative value of eggs will depend on the amount of alternative food sources available, FitzGerald and van Havre, 1987); (2) mating with the male whose nest has been destroyed—a benefit accrued by a single female from the raiding group; and (3) the "spiteful" payoff associated with eliminating potential competitors for your own offspring.

The scant data available suggests that cooperation in nest raiding is an example of byproduct mutualism, as males are able to defend their nests against small groups of females, but are helpless against onslaughts by larger groups. Although only a single female (usually the one who initiates a raid; FitzGerald and van Havre, 1987) will mate with the male whose nest was destroyed, this probability of mating, in addition to the nutritional value of the eggs (no matter how slight) might make being part of a cooperative nest-raiding group worthwhile in an environment that is limited in either good nest sites or high-quality males (a measure of δ; see Chapter 2). Of course, all this rests on the assumption that females in a group do not raid their own nests—an assumption that is supported by FitzGerald and van Havre's (1987) work on three-spined sticklebacks.

3.5   Intraspecific cleaning

Trivers's (1971) seminal paper on reciprocal altruism is full of fascinating possible cases of interspecific cleaning. One of the most interesting, but controversial, examples was cleaning symbioses in fishes (Limbaugh, 1961). Trivers notes:

> The behavior of the host fish is interpreted here to have resulted from natural selection and to be, in fact, beneficial to the host because the cleaner is worth more to it alive than dead. This is because the fish that is cleaned "plans" to return at later dates for more cleanings, and it will be benefited by being able to deal with the same individual. . . . To support the hypothesis that the host is repaid its initial altruism, several pieces of evidence must be presented: that hosts suffer from ectoparasites; that finding a new cleaner may be difficult or dangerous; that if one does not eat one's cleaner, the same individual can be found and used a second time (e.g., that cleaners are site specific); that cleaners live long enough to be used repeatedly by the same host: and if possible, that individual hosts do, in fact, reuse the same cleaner. (1971, p. 41)

Trivers marshals together some of the evidence to support his case, but this example has, nonetheless, come under fire for a number of reasons (Taylor and McGuire, 1988; Gorlick et al., 1978), one of which is that it is difficult to understand the evolution of reciprocity when players do not share a common gene pool (see Poulin, 1993; Poulin and Vickery, 1995, for a game theoretical approach to this question). As we are only considering intraspecific interactions here, I will gently glide by this debate (see Losey, 1972, 1979, 1987, for more) and discuss what little is known on intraspecific cleaning behavior in fish. Intraspecific cleaning behavior is different from interspecific cleaning in that there usually are no "cleaning stations"—specific places where individuals swim in order to be cleaned. Intraspecific cleaning in fishes is, in fact, similar to grooming in mammals, wherein individuals remove ectoparasites from each other's bodies (Fig. 3.4). This behavior has been reported in juvenile salema (*Xenestius californiensis;* Sikkel, 1986), juvenile Panamic sergeant majors (*Abudefduf troschelli;* McCourt and Thomson, 1984), common carp (*Cyprinus carpio;* Soto

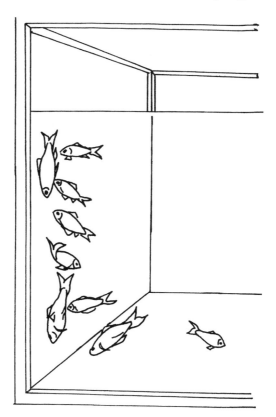

Fig. 3.4. Intraspecific cleaning behavior in *Cyprinus carpio*. The fish in the upper left corner is assuming the typical head-down pose (the request-for-cleaning pose discussed in Soto et al., 1994). Such an individual is approached by another fish and cleaned for a matter of seconds (after Soto et al., 1994).

et. al, 1994), bluegill sunfish (*Lepomis macrochirus;* Abel, 1971; Sulak, 1975), Sumatran barb (*Barbus tetrazona;* Darkhov and Panyuskin, 1988), guppies (*Poecilia reticulata;* Darkhov and Panyuskin, 1988), and sunfish hybrids (French, 1980).

The data available on intraspecific cleaning behavior in fishes make it difficult to tease apart reciprocity and byproduct mutualism as the underlying agent in this case. If the parasites removed provide a significant meal, then cleaning can probably be ascribed to byproduct mutualism. If this is not the case, then it certainly is possible that reciprocity may play a role in cleaning, but no tests have been undertaken to determine the extent to which this is true. However, in many species in which intraspecific cleaning is found, it appears that juveniles are the cleaners and adults are the recipients of this action. This suggests that reciprocity is not likely in such cases, but that kin-related benefits may play a role if juveniles and adults involved in such interactions are related. Future experiments can easily be constructed to examine the roles of these different categories of cooperation. For example, molecular genetic techniques would provide some data on kinship, gut analysis could be used to examine byproduct mutualism, and observational studies using marked individuals could be employed to examine reciprocity.

3.6   Cooperative foraging

In some schooling species of fish, the "harsh environment" that appears to select for cooperative foraging in one species is the territorial defense of another species. That is, territorial defense in species A may cause cooperative group foraging in species B, because *singletons* in species B obtain few, if any, resources (Robertson et al., 1976). Such byproduct mutualism is seen in a number of species of fish (e.g., striped parrotfish—*Scarus criocensis;* Robertson et al., 1976, and surgeonfishes; Vine, 1974; Alevizon, 1976; Montgomery et al., 1980; Foster, 1985a,b; Barlow, 1974) and seems to follow the general pattern described by Foster (1987) for the wrasse *Thalasomma lucasanum.*

In the process of studying group foraging in *T. lucasanum,* Foster (1987) found that the probability of overcoming the territorial, defensive capabilities of male sergeant major damselfish (*Abudefduf troschelli*) and obtaining their embryos (which are attached to the substrate) as prey was directly related to the size of the foraging school (Fig. 3.5). Such groups were never seen when sergeant major damselfish were not nesting, suggesting that facultative grouping evolved as a means of overpowering territory owners. Groups of size < 30 were never able to gain access to such resources, while larger groups (e.g., in the hundreds) were very successful at chasing away damselfish and preying on their embryos. Such attacks could yield up to 250,000 embryos per nest. In the closing sentences of her paper, Foster describes the harsh environment selecting for byproduct mutualism in this system: "These findings imply that effective defense of a high quality resource can favor the development of gregariousness and hence of social behavior among individuals unable to gain access to the resources as solitary foragers" (1987, p. 221).

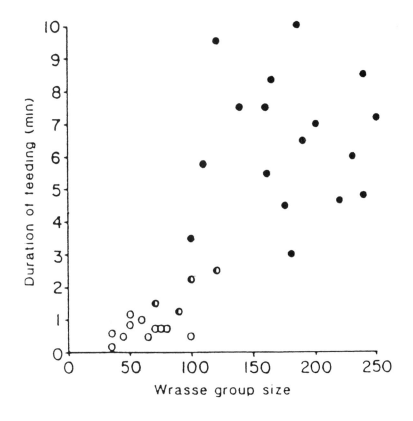

Fig. 3.5. The duration of feeding as a function of group size in the wrasse *Thalasomma lucanum*. Clear circles indicate > 25% of the embryos at a nest were consumed, half-shaded circles 25–50% and shaded circles 100% (after Foster, 1987).

Relatively low density and the consequent difficulty in amassing groups large enough to overpower territory owners seems to have selected against co-operative foraging in another wrasse, *T. bifasciatum* (Foster, 1987).

Anecdotal evidence of even more complex cooperative foraging exists for the yellowtail (*Seriola lalandei;* Schmitt and Strand 1982). Although group foraging in marine piscivores has been shown to facilitate prey capture (Major, 1978), evidence for coordinated hunting tactics among group members is rare. Schmitt and Strand (1982) emphasize this distinction and separate

> cooperative foraging from less complex forms of group hunting behaviors by two criteria: 1) individual predators adopt different, mutually comple-mentary roles during foraging ventures (i.e. there exists a "division of la-bor"), and 2) individuals exercise "temporary constraint" (Wilson, 1975) by not feeding until prey have been rendered more vulnerable. (p. 714)

Though their sample size was very small and the final stages of a hunt were never seen, Schmitt and Strand argue that yellowtails meet this criteria (Fig. 3.6).

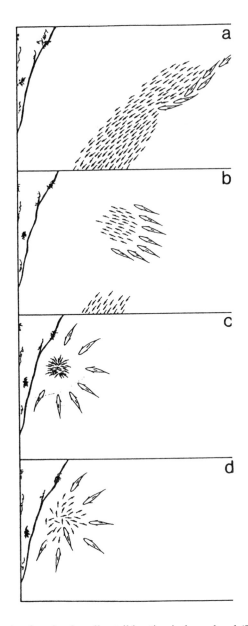

Fig. 3.6. Cooperative foraging in yellowtail hunting jack mackerel (*Trachurus symmetricus*). (a) Yellowtail line up along seaward flank of prey, with lead predators turning into the prey school. (b) A small school of mackerel are split off from their group, and yellowtails have formed a crescent formation used to herd the mackerel toward the shore. (c) The mackerel are pressed against the shore and form a dense aggregation, with yellowtail surrounding their prey. (d) A single yellowtail rushes through the aggregation, scattering mackerel in a radiating fashion (after Schmitt and Strand, 1982).

When attacking large groups of prey, yellowtails act in a coordinated fashion, with individuals apparently taking on specific roles (splitting the school of prey, herding, etc.). Furthermore, similar tactics were used when the same prey species were encountered again, but different, equally complex (and perhaps adaptive) cooperative hunting roles were used when tracking different species of prey. The difference in tactics across prey types indicates remarkable plasticity for such a complex trait as cooperative hunting.

Schmitt and Strand (1982) suggest that cooperative hunting is an example of byproduct mutualism (without using the term itself), as they argue that cooperative hunting in yellowtails evolved as a means of overcoming the difficulties associated with diurnal foraging. Cooperation in yellowtails suggests that although reciprocity is the category most associated with complex cognitive behavior in animals, byproduct mutualism can also lead to an impressive suite of complex actions. The division of labor and the degree of coordination evidenced in yellowtail hunts rivals any example of cognitive complexity seen in the context of reciprocity in fish.

Cooperative foraging of another sort may be occurring in bluegill sunfish (*L. macrochirus;* Mittlebach, 1984; Dugatkin and Wilson, 1992). In a seven-week laboratory experiment, in which all pairwise combinations of (six) fish were tested numerous times, Dugatkin and Wilson (1992) found that bluegills show a strong tendency to associate with group members with whom they have had productive foraging bouts. That is, bluegills use the amount of food they have received when foraging with other fish as a cue in determining with whom to associate. Although this experiment did not examine cooperative foraging per se, the results suggest that bluegills are able to distinguish cooperative from noncooperative foraging partners (see Ranta and Lindstrom, 1990, and Ranta et al., 1992, for evidence that sticklebacks have preferences for foraging partners as well). For other possible examples of cooperative foraging in fish, see Major (1978), Partridge et al. (1983), and Potts (1983).

3.7 Shreckstoff

Karl Von Frisch (1938, 1941) coined the term "Shreckstoff" for the chemicals released by injured fish that appear to elicit fright reactions in conspecifics. Shreckstoff is produced by specialized cells in fish and "general body fluids" do not cause a similar reaction (Smith, 1992). Shreckstoff is most common in the superorder Ostariophysi, but alarm pheromones have been found in perch, gobies, sculpin, cyprinodontiforms, and possibly other taxa as well (see Smith, 1992 and Pfeiffer, 1977, for a complete list of all species of fish which possess alarm pheromones).

A large number of studies have examined Shreckstoff in fish (Smith, 1992), but, unfortunately, the data needed to distinguish between reciprocal altruism, byproduct mutualism, group selection, and kin selection have not yet been gathered. While there is no lack of evidence that conspecifics react to Shreckstoff by performing fright behaviors (see Smith, 1992, for details), it is not known whether the injured individual receives a large benefit as well (Smith, 1986) or whether recipients are typically kin (allowing for kin selection). Furthermore, the possibility that, given an attack, producing Shreckstoff is facultative rather than obligate, has not received the attention it deserves (Smith, 1992). A facultative response would open the door for cooperation via reciprocity or group selection.

Alarm pheromones that are similar in function to those found in fish have been uncovered in sea anemones (Howe and Sheik, 1975), snails (Snyder, 1967), sea urchins (Snyder and Snyder, 1970), tadpoles (Hews, 1988), and rats (Mackay-Sim and Laing, 1982). In addition, possible cases of auditory, tactile, visual, and electric alarm signals in fish are reviewed by Smith (1992).

3.8 Mobbing behavior

Mobbing behavior, wherein numerous potential prey attack or harass their potential predator, has been documented in at least five groups of fish (*L. macrochirus*: Dominey, 1983; butterfly fishes: Motta, 1983; *Stegastes albifasciatus*: Donaldson, 1984; *Pomacentrus coelestis*: Ishihara, 1987; and *Stegastes planifrons*: Helfman, 1989; see Dugatkin and Godin, 1992a for a review). No data are available on the degree of relatedness among mobbers (or between mobbers and nonmobbers in a group), nor is any information known on whether individuals take turns mobbing potential predators. Future experimental work designed to distinguish among categories of cooperation in the context of mobbing need to: (1) use molecular genetic techniques to determine within- and between-group values for r (genes identical by descent), (2) determine whether mobbing is in fact more dangerous than not mobbing, (3) test whether individuals differ in their tendency to mob, and if so, what causes such differences, and (4) examine the affect of a predator's tendency to return to areas in which it has been mobbed. Once this sort of data is gathered, we can begin putting together the puzzle of cooperative mobbing behavior.

3.9 Predator inspection

Pitcher et al. (1986) coined the term "predator inspection" behavior to describe a fish's slow, saltatory movements away from a school and toward a potential predator. Their work on predator inspection was inspired by a study 25 years earlier in which George (1960) examined "predator cone avoidance" in the mosquitofish, *Gambusia affinis*. George (1960) found that some individual mosquitofish moved away from their school and approached a pickerel (*Esox esox*), but avoided the area around the pickerel's mouth (the predator cone). Predator inspection has now been documented in guppies (*P. reticulata*), stickleback (*G. aculeatus*), minnows (*Phoxinus phoxinus*), paradisefish (*Macropodus opercularis),* damselfish (*Stegastes planifrons*), bluegill (*L. macrochirus*), and mosquitofish (George, 1960: see Dugatkin and Godin, 1992a, and Pitcher, 1993, for reviews of predator inspection behavior in fish).

For many reasons (listed below), predator inspection has also become the most widely used experimental system to study reciprocity and the use of the

TFT strategy. Because predator inspection is such a well-studied phenomenon (perhaps the best-studied system we have for cooperation among unrelated individuals), I will devote considerable space to looking at the costs and benefits of this behavior, the possibility that inspectors are trapped in a prisoner's dilemma, the types of strategies used during inspection, and the heated debate over TFT-like behavior during inspection.

3.9.1 The costs and benefits of predator inspection

On the surface, predator inspection seems like a bizarre, even paradoxical behavior. Why should individuals move away from the relative safety of a school, toward a potentially dangerous, hungry predator? In order to answer this question and to begin placing inspection in the context of cooperation, we need to know something about the costs and benefits associated with this behavior. While much work remains to be done, we seem to have at least a qualitative handle on some of the costs and benefits involved with this dangerous activity.

Pitcher et al. (1986) and Pitcher (1993) use such provocative phrases as "Dicing with Death" and "Who Dares Wins" in the titles of their papers on predator inspection, but is it the case that predator inspection behavior is in fact dangerous? Do inspectors truly put themselves at risk, compared with their noninspecting shoal mates? Isn't it possible that inspectors "warn" potential predators that they have been seen, and that any attack would be fruitless, thus making inspectors as least as safe, if not safer, than noninspectors? One potential problem with answering such questions is that the only way to respond definitively is to construct experiments in which inspectors and noninspectors are truly under threat of predation. Ethologists and behavioral ecologists are reluctant to run such experiments because of ethical problems (Huntingford, 1984), however, laboratory studies that use proxies for the level of predation threat (e.g., the number of lunges a predator, behind a partition, makes at inspectors versus noninspectors) cannot sufficiently address the question of mortality threat, forcing the issue of experiments involving true predation.

In my 1992b paper I addressed the costs of predation pressure by constructing ten pools (1 m in diameter), each of which contained a predator, a prey refuge, and a group of six male guppies (each male could be identified by a naturally occurring, unique body-color pattern). Figure 3.7 shows a clear cost to inspection as predation was greatest on the individuals who inspected predators most closely ("high" inspection group), and weakest on those in the "low" inspection group. Godin and Davis (1995a) report data that seem to contradict my data. They found that a predator was more likely to attack and eat a guppy who was not inspecting than an inspecting fish. It is critical, however, to note that noninspecting fish in Godin and Davis's (1995a) study were part of a group in which (at the time) *no one* was inspecting, while the fish in my study were always in groups that contained individuals that differed in their tendencies to inspect. As such, Godin and Davis (1995a) have essentially obtained survival costs of fish in a group when *no one* inspects, and their findings are precisely what we might expect to find if between-group selection was strong for predator inspection. That is, individuals in groups with no cooperators may suffer a

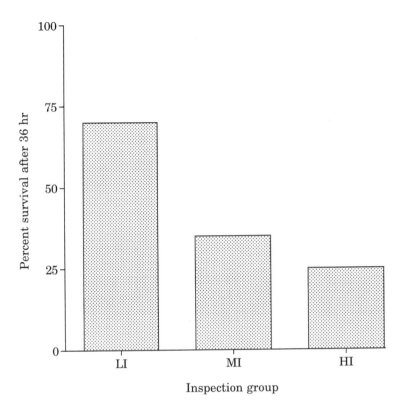

Fig. 3.7. Survival rate after 36 hr as a function of the tendency to inspect predators in the guppy, *Poecilia reticulata*. Sixty fish were initially tested and divided into three groups of 20 each, based on their tendency to inspect predators. HI = high inspection group (approached predator most closely), MI = medium inspection group, and LI = low inspection group.

heavy cost (also see Godin and Davis, 1995a; Milinski and Bolthauser, 1995, for more on this).

Other potential costs to inspection have also been noted. For example, Magurran and Girling (1986) found that inspecting minnows foraged less than noninspecting shoal mates, and Dugatkin and Godin (1992b) found anecdotal evidence for such a cost of foraging in field experiments on predator inspection in guppies. Furthermore, Magurran and Nowak (1991) uncovered a cost unique to female inspectors. Apparently, female guppies who inspect are more susceptible to sneak copulations by males!

What makes these experiments on the costs of inspection particularly interesting in the context of cooperation is that Magurran and Higham (1988) have shown that inspectors somehow transfer the information they receive to noninspectors. In an ingenious experiment involving one-way mirrors, Magurran and Higham (1988) found that once inspecting minnows returned to their school, noninspectors (who could not see the predator) decreased their rate of foraging and increased their rate of skittering (an anti-predator behavioral response). The question then arises, if inspectors are under greater predation pressure than noninspectors, but share the information they obtain during their sorties, how could inspection behavior have evolved in the first place? The answer must be either that the benefits that individual inspectors receive outweigh the cost of predation, or that groups with many inspectors have higher survival rates than groups with few inspectors (the between-group component of selection is greater than the within-group component). Unfortunately, no data are yet available to address the second possibility (but see Godin and Davis, 1995a, and Milinski and Bolthauser, 1995, for some hints), and the discussion now shifts to the benefits of predator inspection.

What sort of information is obtained by inspectors engaged in their dangerous mission? Pitcher et al. (1986) have argued that the potential advantages to predator inspection behavior include: (1) inspection yields information about the potential predator (e.g., is it hunting? how far from the school is it?), (2) inspection informs a predator that it has been seen, and that an attack is therefore unlikely to be successful (Curio, 1978; Curio et al., 1978a), and (3) inspection allows fish to determine if what they see is truly a predator, rather than a non-predatory fish (many species of small fish are sympatric with such large non-predatory species).

A fair amount of evidence exists that inspectors are assessing potential predators and gaining various types of information. One function of inspection behavior seems to be determining whether something that appears to be dangerous is in fact dangerous. For example, Magurran and Girling (1986) and Magurran and Pitcher (1987) have found that if inspectors judge a large fish to be innocuous, they are likely to resume the feeding and courting behavior that they temporarily brought to a halt during inspection. Furthermore, Magurran and Pitcher (1987) found that European minnows were more likely to initiate inspection early in a trial, rather than later, when the predator was moving and stalking the minnows. That is, one function of inspection in minnows appears to be to assess whether a predator is in fact in "hunting mode." An extreme example of this is that fish are less likely to inspect a predator after it has been observed making a kill or in possession of a dead prey individual (Csanyi et al., 1985; Csanyi, 1985; Magurran and Pitcher, 1987). Inspectors appear to make even finer-level distinctions. For example, Licht (1989) has found that guppies are able to distinguish between hungry and satiated predators.

Another putative benefit of inspection is deterring a predator from attacking. Even though inspection has been shown to be a dangerous activity in some contexts (Dugatkin, 1992b), if the benefits of deterring predators in other contexts are great enough, inspection could evolve through classic natural selection at the individual level. Only one study to date has addressed the issue of whether inspection behavior deters a predator from attacking. Magurran (1990) found that inspection by minnows did in fact deter an attack from a pike predator. It is important to note that the fact that inspection can deter attack is not contradictory with the evidence that inspectors are under greater predation threat than noninspectors. Although untested, it may simply be that inspection often deters attack, *but if an attack occurs,* inspectors are preferentially taken. One last benefit that may play a role in predator inspection by males is female mate choice. Godin and Dugatkin's (1996) work suggests that one benefit of inspecting may be that inspectors are more attractive to females, and hence gain an advantage with respect to mating opportunities.

Taking into account many of the costs and benefits described above, I (1990) and Dugatkin and Godin (1992a) developed a game theoretic model that examines the evolution of predator inspection in fish. For various shoal sizes, this model yields the equilibrial frequency of inspectors (and consequently noninspectors), the optimal distance an inspector should inspect, and the predicted distribution of inspector group sizes. A qualitative test of the model using data from guppies from Ramdeen's Stream (a tributary of the Arima River) in Trinidad suggests that both the costs and benefits of predator inspection are intermediate (not extreme in either direction). It is worth noting that compared with other tributaries in the Arima River, Ramdeen's stream appears to have an intermediate number of piscine predators.

Now that we have a good understanding of the costs and benefits of inspection, I can move on to an issue nearer and dearer both to my heart and to the subject of this book: cooperation.

3.9.2   Are inspectors trapped in the iterated prisoner's dilemma?

Even before most of the cost and benefit work on predator inspection was initiated, Milinski (1987) suggested that pairs of inspecting fish may be trapped in a prisoner's dilemma (Fig. 3.8). Recall that the prisoner's dilemma requires $T > R > P > S$ and $R > (S+T)/2$ (Fig. 1.1). Although his experiments did not quantify the prisoner's dilemma payoff matrix, enough evidence about the payoffs of inspection are now available to address this question in some detail.

Is $T > R$? Milinski (1987) argued that the trailing fish in a pair of inspectors can obtain information about the potential danger by watching what happens to the lead fish, and hence $T > R$. That is, the trailing fish is better off staying back and observing whether the lead fish gets attacked (and obtaining T, the temptation to cheat), than swimming up beside the lead fish (and obtaining R, the reward for mutual cooperation). Evidence that inspectors are more likely to get eaten the closer they approach a predator (Dugatkin, 1992b), coupled with the evidence that inspectors transfer the information they receive, provides some empirical support that T is in fact greater than R for predator inspection. In addition to the empirical support that $T > R$, there is a logical argument put forth that this inequality indeed holds true. In a response to a critique of his work by Lazarus and Metcalfe (1990), Milinski argues:

> If T were not greater than R, then after the first fish had moved towards the predator, not only a second one, but a third, a fourth and so on should follow until all the minnows or sticklebacks present in the tank were close to the predator. (Milinski, 1990, p. 990)

Data from minnows (Magurran and Pitcher, 1987) and guppies (Dugatkin and Godin, 1992b) indicate that all shoal members do not inspect a predator together. For example, field experiments with groups of six guppies show that singletons and pairs were the most common inspecting group sizes (Dugatkin and Godin, 1992b; see Turner and Robinson, 1992; Milinski, 1992, for more on the frequency of inspector group sizes).

The question of whether or not $T > R$ has direct consequences on whether inspection is better explained by reciprocity or byproduct mutualism. If T is greater than R, and if other conditions of the prisoner's dilemma hold true, then inspectors are trapped in this game and one might expect reciprocity to evolve under some circumstances. However, if $R > T$, then a big share of the temptation to cheat is removed; players recieve greater benefits by inspecting than they do by refraining, and inspection is best explained by byproduct mutualism (see Connor, 1996; Dugatkin, 1996; Milinski, 1996, for more on this, and Stephens et al., 1997b, for data that inspection in mosquitofish is best explained by byproduct mutualism).

Is $R > P$? The second requirement for a matrix to be a prisoner's dilemma is that the payoff to mutual cooperation (R) be greater than the payoff to mutual

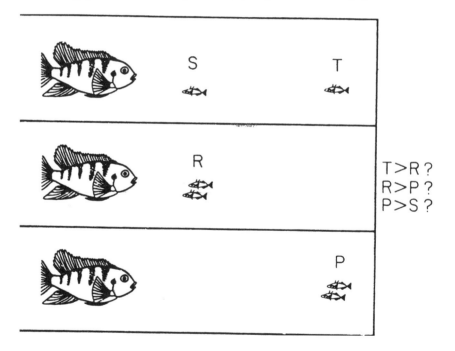

Fig. 3.8. Predator inspection behavior in pairs of fish as a possible prisoner's dilemma game. When one fish alone inspects, it receives the "sucker's" payoff (S), with its partner obtaining the "temptation to cheat" payoff (T). When both inspect, each receive the reward for mutual cooperation (R), and when neither inspects, they receive the punishment for mutual defection (P) (after Milinski, 1990).

defection (*P*). A line of logical arguments similar to that invoked above suggests that *R* is in fact greater than *P*. That is, if *P* > *R* then it would never pay for any fish to inspect a predator and the phenomenon of inspection would be rare and maladaptive when it occurred. Given that inspection is a well-known phenomenon, this is unlikely to be the case. Although this inequality has not been investigated directly, indirect experimental evidence suggests that *R* > *P*. For example, inspectors, in some yet unknown fashion, transfer information to noninspectors (Pitcher et al., 1986; Magurran and Pitcher, 1987; Magurran and Higham, 1988). Noninspectors subsequently decrease foraging and increase anti-predator behaviors, if the information they receive suggests that the potential predator is a real danger. Similarly, Godin and Smith (1989) have shown that fish that are "confused" by potential foraging opportunities are more likely to be taken by predators. These experiments, though providing only indirect evidence, suggest that a school of fish with no inspectors (which is what we would predict if *P* > *R*) would be under severe predation threat.

Is P > *S?* For the payoffs of predator inspection to qualify as a prisoner's dilemma, it must also be true that the payoff to mutual defection (*P*) must be greater than that associated with inspecting as a singleton (*S*). Although it appears to be the case that having no inspectors in a shoal is dangerous, perhaps the most dangerous situation for a single fish to be in is to be the lone inspector in a group (Pitcher's 1993 "sacrificial lamb"). Evidence from a number of experiments (e.g., Milinski, 1977a,b) indicates that single fish (stragglers) suffer very high rates of predation, suggesting that *P* > *S*.

Is R > (*T* + *S*/2)? It has often been argued that for a game to be a prisoner's dilemma, it must also be true that the payoff for mutual cooperation (*R*) must be greater than the payoff for alternating between cooperation and defection ((*S* + *T*/2)). Milinski (1987) argues that this last inequality is also met by predator inspection: "*R* > (*S* + *T*/2) should also be fulfilled. A group of two fish may detect an impending attack earlier which may be less frequently successful anyway because of the confusion effect of two prey compared with one" (p. 435).

   Although the jury is still out, the experimental evidence provided above, coupled with the logical arguments for these inequalities, suggest that when inspection occurs in pairs, the requirements of the prisoner's dilemma are often met.

3.9.3   Do inspectors use the Tit for Tat strategy?

Given that the payoffs of predator inspection approximate a prisoner's dilemma, Milinski (1987) and I (1988) asked whether fish inspecting a predator in pairs use the TFT strategy, as theory would predict. Using mirrors of different lengths, and placed at different angles, Milinski (1987) devised an elegant experiment to examine whether inspecting fish use the TFT strategy.

   In Milinski's 1987 experiment, single sticklebacks were tested in one of two treatments. The "cooperation" treatment had a mirror running the entire course of the stickleback's tank, from the refuge at one end to the tank containing the predator at the other end (Fig. 3.9). This mirror simulated a second stickleback that stayed by the side of the subject when it inspected the predator (which was itself in a separate, but adjacent tank). That is, the mirror image always cooperated with the stickleback when it inspected the predator. In the "defection" treatment, a mirror whose length was only half that of the sticklebacks' tank was used. This mirror was placed at a 32° angle from the stickleback refuge (Fig. 3.9). As such, this mirror simulated a co-inspector that stayed by a subject's side when it was near the refuge and far from the predator, but lagged further and further behind during an inspection by the subject, until it defected (could not be seen by the subject). In 1988 I used an identical protocol to examine whether guppies were using the TFT strategy during inspection.

Milinski (1987) and I (1988) used two lines of evidence to argue that stickle-

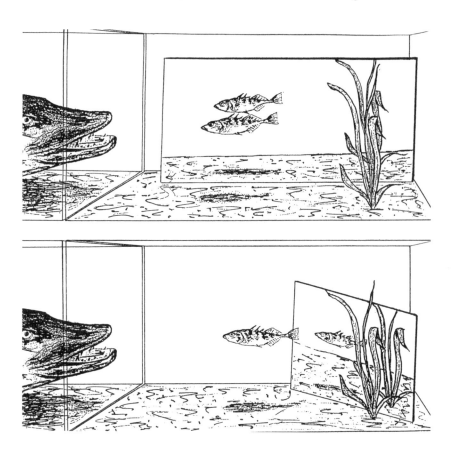

Fig. 3.9. Predator inspection in sticklebacks. In the top panel, a single fish observes its mirror image remain by its side as it inspects the predator (the "cooperating mirror" treatment). In the lower panel, the inspector sees its image lag behind and eventually disappear during an inspection visit (the "defecting mirror" treatment; the mirror is angled at 32°). (Drawing provided by Claus Wedekind)

backs and guppies, respectively, used TFT during predator inspection visits: (1) Fish in the cooperating mirror treatment were much more likely to be found in sections of the tank closest to the predator, while fish in the defecting mirror treatment were more likely to be found in the sections of the tank furthest from the predator. That is, when a co-inspector stayed by its side during an inspection, an individual fish was willing to continue inspecting, otherwise it was not. As such, inspectors employed the "do what your opponent did the last time" part of the TFT strategy. (2) Inspectors appeared to "forgive" prior moves of defection if they were followed by moves of cooperation, as dictated by the TFT strategy. This was examined as follows: Fish in each mirror treatment were divided into two groups, based on how closely they approached a predator *during the first half of a trial.* Within each treatment, the fish that inspected closer to the predator were labeled "bold," and the other half were labeled "cautious." The data were then analyzed to determine whether bold and cautious fish were consistent in their behavior during a trial—that is, were their average positions similar during the first and second half of a trial? TFT would predict that bold and cautious fish in the cooperating mirror treatment should be consistent, as their image is always by their side. TFT would, however, predict that in the defecting mirror treatment, only cautious fish should be consistent.

The argument goes as follows: in the defecting mirror treatment, bold fish should recognize that the co-inspector (their image) was not staying by their side early on, and thus should not approach the predator as closely in the second half of the trial. Cautious fish in the defecting mirror, however, typically stayed far enough away from the predator that their mirror image (even though they are in the defecting mirror treatment) was relatively close to them. Data support this argument, as bold and cautious fish in the cooperation treatment, as well as cautious fish in the defection treatment, were consistent, while bold defectors were not. Furthermore, all cooperators and cautious defectors were, on average, closer to the predator at the end of a trial than at the start, but this was not true for bold fish in the defection treatment, whose average position did not change over the course of a trial. This suggests that bold fish in the defecting treatment were "forgiving"—that is, when they went out too far and their image disappeared, they turned back (retaliated). Once they retaliated and swam back away from the predator, however, their image was once again close to them and they began inspecting once more. That is, they forgave prior moves of defection when partners cooperated in the present.

Using all of the above information, Milinski (1987) and I (1988) argued that inspector's behavior approximated that of the TFT strategy.

3.9.4   The fire after the storm—critiques of the Tit For Tat inspection experiments and replies

The suggestion that individuals use the TFT strategy did not go unchallenged. A series of five critiques and responses on TFT and predator inspection appeared in the journal *Animal Behaviour.* This exchange, while sometimes appearing endless, served a useful purpose, in that it spurred studies that would not have been undertaken in its absence. Next, I review some of the critiques and responses, and try to examine how they better help us understand the evolution of cooperation in the context of predator inspection.

3.9.4.1 "The no-predator control experiment" critique

Both Lazarus and Metcalfe (1990) and Masters and Waite (1990) argue that since a no-predator control experiment was not run in either Milinski's (1987) or my (1988) original experiments, a more parsimonious explanation than TFT can be constructed for the fish's behavior. That is, without knowing how the fish would behave in a control experiment in which no predator was present in the "predators tank," Milinski's and my results just indicate that fish prefer to have another conspecific near them, regardless of the presence of a predator. As such, no need exists for more complex explanations involving TFT.

In their critique, Masters and Waite (1990) went so far as to run a no-predator control experiment themselves (although for some reason they did not repeat the predator-present treatment). They found that when no predator was present, fish still approached the "predator tank" more closely in the cooperating mirror treatment, as in both Milinski (1987) and Dugatkin (1988). Masters and Waite conclude:

> [O]ur results suggest that the concepts of predator inspection and cooperation need not be invoked in the experiments of Milinski and Dugatkin. Their results may simply reflect the tendency of shoaling fish to aggregate. (1990, p. 604)

Fortunately, they follow this up with a suggestion for a test that would allow one to distinguish TFT from their simpler "shoaling" hypothesis:

> If angled-mirror designs are used in the future, tests should be done with and without the predator. Evidence of tit-for-tat-like behaviour would be a predator-mirror interaction, i.e. closer approach to the predator when the mirror is parallel than would be expected due simply to the tendency to shoal. (p. 604)

I (1991a) ran just such an experiment and indeed found the predator-mirror interaction that Masters and Waite (1990) argued would be evidence of a TFT-like strategy on the part of inspectors.

3.9.4.2 "The payoff matrix" critique

According to Lazarus and Metcalfe (1990) and Connor (1996), the payoffs for predator inspection may not meet the demands of the prisoner's dilemma. TFT is a strategy that works when players are trapped in a PD, and hence the critics argue that since inspectors aren't in a prisoner's dilemma, they are not using TFT. Since this critique revolves around whether the payoffs for inspection satisfy the prisoner's dilemma, I direct the reader to the prior section.

3.9.4.3 The "lack of individual recognition" critique

Both Lazarus and Metcalfe (1990) and Masters and Waite (1990) have argued that the behavior displayed by inspecting fish can be explained by a simple rule that instructs players to stay within a certain distance of a co-inspector, regardless of who they are or how they have behaved, and hence there is no need to infer the use of TFT. That is, in the case of inspection, TFT requires an ability to recognize co-inspectors and remember their tendency to inspect or not inspect in the past. Such evidence of individual recognition of co-inspectors would be stronger evidence for TFT than that provided by the mirror experiments (which were not designed to address the issue of individual recognition; but see Connor, 1996; Dugatkin, 1996; Milinski, 1996).

Milinski et al. (1990a,b) and Dugatkin and Alfieri (1991a,b) have addressed these questions in a series of experiments designed to examine whether inspectors recognize their partners, remember their prior interactions with such partners, and temper current behavior based on the actions of their partners in the past. Milinski et al. (1990a) and Dugatkin and Alfieri (1991a,b) found that individual inspectors remembered the behavior of a particular partner and were much less likely to cooperate with a partner that had cheated in the past. Milinski et al. (1990b) attacked the question somewhat differently. Given that inspectors are likely to have the same shoalmates, Milinski et al. (1990b) asked whether individuals had "favorite" partners. They found that in groups of four sticklebacks, individuals often inspected in set pairs, implying that inspectors are able to recognize one another and choose potential partners based on some criteria. In sum, the work on individual recognition suggests that fish recognize one another. This is an important point, because it suggests that models which demonstrate that the dynamics of inspection can be explained by simple "automaton-like" rules on the part of school members (Stephens, 1997a), are insufficient to explain behavior when experimental evidence for individual recognition and context-dependent behavior exists within the predator inspection literature. Furthermore, we must not lose sight of the importance of score-keeping, even if it isn't always in the context of TFT.

3.9.4.4 The "no-live partner" critique

Perhaps the most basic critique of the inspection experiments was that although mirrors allowed for control of a "partner's" behavior, they did not allow the behavioral dynamics among two fish to be observed, thus preventing us from obtaining critical information on whether inspectors displayed the three characteristics of TFT: start off cooperatively (be nice), cheat on cheaters (show retaliation), and don't hold grudges (be forgiving) (Lazarus and Metcalfe, 1990; Masters and Waite, 1990; Reboreda and Kacelnik, 1990).

I (1991b) attempted to quantify these characteristics when testing pairs of fish inspecting a predator over the course of a five-minute trial. Given that a game of inspection may begin when one fish commences inspection, niceness was measured by comparing the latency between when the first and second fish began inspecting in a predator "present" and predator "absent" treatment. Significant differences across the predator treatments were considered evidence of nice behavior. That is, once the game started and one fish began inspecting, nice behavior would dictate that its partner start off cooperating (approaching) when the predator was present, but not when the predator tank was presented empty. Holding absolute position in the tank constant, retaliation was defined as the case in which the lead fish was more likely to stop and end an inspection than was the trailing fish. Forgiving was measured by examining the tendency of a lead fish that just retaliated to begin inspecting again, if its partner initiated a new round of inspection.

I (1991a) found evidence for all three characteristics during guppy inspection behavior. Niceness, retaliation, and forgiveness (defined somewhat differently) were also found when a similar experiment was run over a longer time course (Dugatkin and Alfieri, 1991b). Furthermore, guppies from populations that are not sympatric with large piscine predators do not display any of the characteristics of TFT when they inspect predators (Dugatkin and Alfieri, 1992), suggesting that natural selection has shaped the TFT behavior found during predator inspection (Endler, 1986).

3.9.5 Predator inspection wrap-up

Although probably biased, I believe the evidence, overall, indicates that predator inspection is a good system in which to study reciprocity, both because of the dynamics of the behavior and the structure of the payoff matrix (i.e., a prisoner's dilemma). The experimental evidence also suggests that inspectors are in fact using something akin to TFT. The vociferous nature of the debate surrounding the use or nonuse of TFT served a useful purpose, in that it spurred on experimental work that allowed us the best understanding yet of the dynamics of the TFT strategy in a given ecological and behavioral context.

3.10 Summary

The evidence for cooperation in teleost fishes was reviewed. Eight areas—egg trading, helpers-at-the-nest, nest raiding, cleaning behavior, foraging, alarm chemicals, mobbing, and predator inspection—were outlined, and the role of byproduct mutualism, reciprocity, group selection, and kin selection was examined to better understand cooperation within each of these areas. All four categories were evident (though not in equal proportions) within the fishes, and this taxa, for many reasons, should continue to provide interesting insight into the evolution of cooperation.

4

Cooperation in birds

4.1 Introduction

Ornithologists, perhaps more than any other "ologists" in biology, are likely to have found their calling early in life, and all seem to have a picture of themselves at age three, with birding binoculars in hand. As such, it should not be surprising that the literature on social behavior in birds is large indeed. In this chapter, I will discuss a number of different types of avian cooperative behavior and, as always, tie it back to theory, whenever possible. In particular we will look at cooperation in the context of territoriality, hunting, food sharing, alarm calls, and mobbing.

One gaping hole in the list above is a discussion of cooperative and communal breeding. As much as I would like to cover this topic, any attempt to address this issue as a part of one chapter in a single book is folly, and so I ask for the reader's indulgence, when I simply cite a list of books and edited volumes on this topic. These books alone would keep most behavioral ecologists interested in the evolution of cooperation busy and entertained for a long period of time. I will not even attempt to provide a partial list of articles on this topic. New papers on cooperative and communal breeding are consistently appearing in major behavior and evolution journals, and as far as we can see, just about every issue of every ornithology journal has a new paper on cooperative/communal breeding. Given all that, I refer the reader interested in starting to learn about cooperative and communal breeding in birds to the following works: Emlen, 1984, 1991; Woolfendon and FitzPatrick, 1984; Brown, 1987; Koenig and Mumme, 1987; Skutch, 1987; Stacey and Koenig, 1990; Bertram, 1992; Davies, 1992; and Marzluff and Balda, 1992.

4.2 Territoriality and cooperation

4.2.1 "Dear enemies" and defensive coalitions

Territoriality is very common among birds (e.g., Wilson, 1975; Brown, 1975), and it is easy to slip into the habit of thinking of territorial neighbors as being

71

in a strictly competitive mode with one another. Territoriality, however, may foster cooperative bonds between neighbors, and may even lead to what has been called the "dear enemy" effect. Fisher (1954) introduced this idea, as follows:

> [T]he effect of holding territory by common passerines is to create "neigh-bourhoods" of individuals which are masters of their own definite and lim-ited property, but which are bound firmly and socially, to their next door neighbours by what in human terms would be described as a dear enemy or rival friendly situation, but which in bird terms should more safely be described as mutual stimulation. (p. 73)

Getty (1987) further adds that:

> Even if neighbors are locked in severe competition among themselves, they often have common enemies and mutual interests that add a cooperative dimension to a competitive situation. (p. 327)

Krebs (1982) portrays the dear-enemy phenomenon as one in which territo-rial neighbors are at first involved in a costly battle over setting up territory boundaries. Once these boundaries are established, however, neither party stands to gain by having to "renegotiate" such boundaries with a new neighbor, and so a new cooperative, nonaggressive component is added to their relation-ship—producing dear enemies (see Falls, 1982; Ydenberg et al., 1988, for re-views). In this light, the dear-enemy strategy has been defined as:

> (1) not challenging your neighbor for portions of its territory or pilfering from its mutually recognized domain, and (2) not being particularly vigilant against your neighbor, on the assumption that it can be trusted not to chal-lenge, or pilfer. Getty (1987, p. 332)

Getty (1987) employs the prisoner's dilemma and the economic metaphor of the "oligopoly" to model: (1) the dear-enemy phenomena, and (2) active coali-tions among neighbors to inhibit third party "floaters" from challenging either territory holder. In his dear-enemy model, Getty (1987) demonstrates that the payoffs associated with dear enemies fit the requirements of the prisoner's di-lemma (see Chapter 2), and proceeds to examine the conditions under which the Tit for Tat strategy is a solution to this game (but see Ydenberg et al., 1988, for the view that the dear-enemy idea can best be explained using the asymmet-ric war of attrition developed by Parker and Rubenstein, 1981). In the "defen-sive coalition" game, Getty tackles the question of what conditions favor ac-tively chasing intruders (floaters) from the territory of neighbors (Getty, 1987). Again using the prisoner's dilemma game, he examines two strategies: "Be a dear enemy, but don't help your neighbor," and "Be a dear enemy and help your neighbor." Results indicate that TFT can be an ESS to this game for reasonable parameter values.

Hooded warblers *(Wilsonia citrina)* are a good system to test Getty's ideas, as individuals in this species recognize their neighbors and are involved in many interactions with them over a long, but indeterminate period of time (Godard, 1991). As such, Godard (1993) examined whether hooded warblers used a TFT or TFT-like strategy in the context of neighbor interactions. Using a taped playback system, Godard (1993) simulated an "intrusion" deep into male A's territory. Intrusions were either playbacks of a territorial neighbor (male B) or a stranger. A's responses to a playback of B's song were examined before and after such an intrusion. Using seven different behavioral measures, Godard (1993) found that male territory owners were much more aggressive to neighbors after a treatment if the neighbor's song was played deep within that male's territory. In addition to providing further evidence that warblers recognize their neighbors, this study clearly demonstrates the "retaliatory" component of TFT, as well as "niceness" (subjects did not respond aggressively to a neighbor's song before intrusions). While "forgiving" was not experimentally examined in this study, Godard (1993) noted that heightened aggression among territorial neighbors was evident on the day of a neighbor's intrusion, but not on subsequent days, suggesting the possibility that warblers forgive neighbors if they do not repeat the same offense. Of course, an equally plausible argument is that hooded warblers simply forget about intrusion over the course of time (see Healy, 1992, for more on optimal forgetting).

4.2.2   Owners, satellites, and territory defense in pied wagtails

Davies and Houston (1981) provide a fascinating account of pied wagtail (*Motacilla alba*) behavior. Their work provides behavioral ecologists with much food for thought with respect to the economics of territoriality, the feasibility of simple optimal foraging models, and the evolution of cooperative territory defense. Pied wagtails defend riverside winter territories, foraging on insects that wash up on the banks (Davies, 1976). This food source is "renewable" in the sense that after a period of time, prey abundance on a territory again increases after being depleted, and Davies and Houston (1981) demonstrate how wagtails systematically search their territories, providing time for depleted resources to again increase.

Table 4.1   The distribution of territory defense among owners and satellites (after Davies and Houston, 1981).

No. defenses by		Percent defense
Owner	Satellite	by satellite
13	5	27.8
2	1	33.3
8	2	20.0
6	6	50.0
44	22	33.3
8	7	46.7
13	15	53.6
94	58	38.1

The story is more complex, however, because intruders often land on such territories. At times such intruders are tolerated (at which point they become "satellites") and at times they are aggressively chased off a territory. What differentiates these situations? When answering this, it is important to note that a territory is worth more to a pied wagtail owner than to an intruder, because intruders do not know which areas have recently been "cropped", while owners do. If an owner permits an intruder to stay, it loses some foraging-related benefits, but gains assistance in territory defense (Table 4.1; Charnov et al., 1976; Davies and Houston, 1981). Davies and Houston (1981) devise a model, the "owner's dilemma," and use this model to predict when owners should allow satellites on their territories. As predicted, Davies and Houston (1981) found:

> On days of high food abundance, when intruder pressure is greatest, owners tolerate satellites . . . owners increase their feeding rate by this association, because the benefits gained through help with defence outweigh the costs incurred through sharing the food supply with another bird. On days of low food abundance, when an owner would have a higher feeding rate by being alone, it evicts the satellite from the territory. (p. 157)

This is a particularly interesting case of conditional cooperation because it suggests that the costs and benefits of byproduct mutualism can vary depending on territory-holding status, and that intruders, who have no foraging territories of their own, always face a harsh environment with respect to food allocation and are willing to pay the costs of territory defense. When food becomes more sparse for owners, however, they do not allow intruders in. The harsh environment that allows the presence of satellites on territories is not the amount of food on a territory, but the amount of "pressure" from other conspecifics. So, intruders are always willing to undertake cooperative territory defense (in exchange) for food, but the costs and benefits of byproduct mutualism do not always select for owners to permit this cooperative act (Brown, 1982).

4.2.3 Pukekos as prisoners

It is easy to forget that one robust solution to the prisoner's dilemma game is to always cheat. To illustrate this, consider Craig's (1984) work on communally breeding pukekos *(Porphyrio porphyrio)*. Craig (1984) formalizes the pukeko's dilemma as follows (Table 4.2): Breeding males can remain the solitary male on their territory, and if neighboring individuals both cooperate in this fashion (i.e., do not take on any other males on their territory), they each receive a reward for their mutual cooperation. However, a pukeko can also opt to cheat and accept another male, who may assist in territory defense, and, in so doing, such a male becomes a communal breeder. Since most defensive interactions occur with neighbors, it is not surprising that a comparison of territories which persisted for more than a few months showed that those defended by one male had shorter boundaries. (Craig, 1984, p. 148). That is, the temptation to cheat *(T)* can be quite large in this system. However, it also turns out that the payoff for mutual cooperation *(R)* is greater than that for mutual defection *(P)*, as

Table 4.2. A possible payoff for solitary and communal breeding pukeko males (after Craig, 1984).

	Pukeko B	
	Solitary (in pair)	Communal
Solitary (in pair)	R = 2.0 Reward for both males remaining in pairs	S ≅ 0 Payoff for solitary male if neighbor becomes communal
Pukeko A		
Communal	T > 2 (?) Temptation to become communal if neighbor does not	P = 0.6 Payoff if both males become communal

breeding pairs produced more offspring per adult than individuals breeding communally. Pukekos trapped in this dilemma have opted for a strategy of defection rather than conditional cooperation, "In most examples, the unit losing ground admitted an extra adult male to counter this imbalance on the number of defenders." (Craig, 1984, p. 148).

Why the pukeko has opted to defect as a solution to this game, even though individuals interact many times (i.e., w is quite high) remains unclear, but future work on the value of matrix entries may shed light on this question. For example, it may be the case that even though the average number of offspring/adults on communal territories is less than that on breeding-pair territories, if a male who allows another individual onto its territory maintains priority access to the female, its reproductive success may be quite high. If this was the case, mutual defection would have a greater payoff associated with it than mutual cooperation, making any dilemma disappear.

4.2.4 Tree swallows, territories, and Tit for Tat

In the first controlled experiment looking at TFT after Axelrod and Hamilton's model (1981), Lombardo (1985) tackled the question of whether parental breeders and nonparental "floaters" in tree swallows *(Tachycineta bicolor)* were trapped in a prisoner's dilemma, and, if so, what sorts of strategies they used. As was the case for warblers and the dear-enemy strategy, tree swallows are ideal candidates for strategies invoking reciprocity because they (1) breed in aggregations, allowing repeated interactions, (2) can recognize each other as individuals, and (3) display intraspecific aggression (Lombardo, 1985).

Before moving on to Lombardo's experiment, let us take a moment to see just how breeders and nonparental floaters may be involved in a game that resembles the prisoner's dilemma. One might expect breeding individuals to always chase away floaters who are attempting to learn about future nest sites

(Leffelaar and Robertson, 1984; Stutchberry and Robertson, 1985, 1987), but they do not (Lombardo, 1985); the situation is more complex, as allowing floaters onto a territory has both costs and benefits. By allowing floaters onto a territory, breeders obtain potential nest site defenders, but if floaters act aggressively, a nest owner's vulnerable offspring are the most likely victims (see Lombardo, 1990, for a list of the potential damage nonbreeders can cause). Lombardo phrases this in game theory terminology and notes:

> Parents showing restraint would allow nonbreeders to visit their nest box unhindered; an act of defection would prevent nonbreeders' visits by vigorous nest defense. For nonbreeders, an act of restraint would be a benign visit or one in which they joined in the mobbing of a predator; an act of defection would be to behave in such a way as to lower parental reproductive success. (1985, p. 1364)

As such it seems that breeders and nonbreeders would each do better if they defected and the other cooperated, but mutual cooperation appears to yield a higher payoff than mutual defection—the game then, appears to be a prisoner's dilemma in which restraint is synonymous with cooperation. Such games—where cooperation is equated with restraint (i.e., doing nothing)—are notoriously difficult to examine, as it is very hard to measure how many times an animal doesn't do something.

Lombardo (1985) examined whether players in this game used TFT by simulating an act of defection on the part of nonbreeders. Nonbreeder male models were placed on a breeder's territory and parental responses were recorded. Parents did not typically respond aggressively to models at this point. Subsequently, two live chicks were removed from the nest and replaced by two dead chicks. This removal was intended to simulate massive defection by floaters. A control was run in which two chicks were banded, rather than replaced by dead chicks. When parents found two of their chicks dead, they retaliated aggressively, chasing nonbreeders off their territory. Interestingly, live nonbreeders who were present were *not* chased if their appearance precluded them as the killers. In addition, aggressive responses to model nonbreeders and live nonbreeders decreased with time. Lombardo interprets this as evidence for forgiving behavior on the part of breeders. Combining this with the fact that breeders show initial restraint, and then subsequently retaliate against the death of a chick, Lombardo (1985) argues that the evidence supports the claim that breeders are using TFT.

In order for TFT to persist as the solution to this game, mutual cooperation payoff (R) must be quite large, to make up for the huge loss when floaters cheat. In fact, in response to a critique by Koenig (1988), Lombardo (1990) now argues that R is large, so large that Lombardo (1990) now claims that $R > T$, and that the restraint seen in tree swallows may best be thought of as byproduct mutualism. If R is large in this game, however, it is not clear why TFT should evolve in such a system, as the payoff matrix is now such that nonbreeders should not cheat in the first place.

4.3 Cooperative hunting

4.3.1 Harris' hawks

Group hunting behavior has been reported in many species of raptors. Unfortunately, much of this evidence is anecdotal (see Bent, 1938; Cade, 1982), and little if any data are available on successful versus unsuccessful hunts, let alone intake rate, making it difficult to assess whether these are truly cases of cooperation (Ellis et al., 1993). An exception to this is Bednarz's (1988) work on cooperative hunting in Harris' hawks *(Parabuteo unicinctus)* in New Mexico. All groups examined in Bednarz's study contained a breeding male and female and, depending on group size, zero to two adult plumaged "auxiliaries" (who were raised on the breeding pairs territory) and zero to three immature hawks, reared during the most recent breeding attempt. Groups were very cohesive, making it possible to track them by following key members, and average group size at the 29 kills recorded was 4.4 individuals.

Harris' hawk attacks are quite coordinated and individuals involved in such hunts use at least three different cooperative hunting tactics:

> The most common tactic may be described as a surprise pounce (seven kills) involving several hawks coming from different directions, and converging on a rabbit from far away. . . . When a rabbit found temporary refuge or cover, a flush-and-ambush strategy was employed. Here, the hawks surrounded and alertly watched the location where the quarry disappeared while one or possibly two hawks attempted to penetrate the cover. When the rabbit flushed, one or more of the perched birds pounced and made the kill. . . . The final tactic used "relay attack" was the least common (2 of 13). This technique involved a nearly constant chase of a rabbit for several minutes while the "lead" was alternated among party members. (Bednarz, 1988, p. 1527)

A quite impressive array of hunting tactics indeed!

Although Bednarz (1988) almost never observed a solo hawk hunting, a critical question is still whether individual success rate increases with hunting-group size. Figure 4.1 clearly indicates that it does, as both the number of prey items successfully captured increases with group size, as does the average kilocaries per individual (see Montevecchi, 1979 for a similar relationship between group size and hunting success in the raven, *Corvus corax*). In addition, the benefits of increasing group size may asymptote at about five individuals/group—the most common group size observed in the nonbreeding season. Bednarz and Ligon (1988) also found that group living was not related to territory quality, but rather to the increased food intake of cooperative hunting, suggesting a strong role for byproduct mutualism and kinship (all group members are related) in this system.

4.3.2 Falcons

Hector (1986) examined group foraging in Apomado falcons *(Falco femoralis)*, one of the "cooperative-hunting falcons" *(F. chicquera, F. berigora, F. jugger,* and

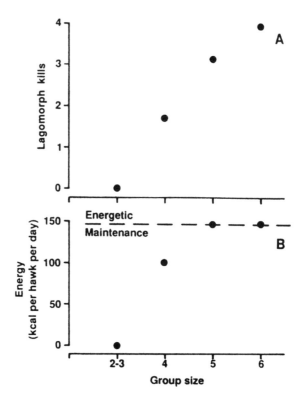

Fig. 4.1. (a) Capture rates by Harris' hawks as a function of hunting group size. (b) Estimated average energy available to hawks as a function of hunting group size. The dashed line represents estimated energy needed for maintenance (after Bednarz, 1988).

F. biarmicus; Hector, 1986). Although no data on the average intake rate of individuals in different sized groups is presented, the increased success rate of pairs of falcons hunting in unison indicates that cooperative hunting may be a real phenomena in this species. Hector (1986) sums up the evidence for cooperation:

> two falcons chased the same prey individual when pursuing flocks; females and males tended to adopt distinct roles in hunts; a simple call seemed to promote joint pursuit of common prey; food was shared by adults; and there was some indication . . . that birds were monitoring each other's movements during hunts. (p. 253)

For a review on cooperative hunting in raptors, see Ellis et al. (1993), and for a more general review of this subject, see Packer and Rutton (1988).

4.4 Food calls, information, and food sharing

Newman and Caraco (1989) developed a game theory model for the evolution of cooperative food calling (similar to Caraco and Brown's 1986 model). Newman and Caraco's model incorporates both the risk of starvation and the risk of predation on the part of callers and noncallers. In addition, they consider both the mean and variance associated with the size of the food patches. Depending on the situation, calling may evolve via byproduct mutualism, via reciprocity, or not at all. Byproduct mutualism represents the solution to the game when the environment is harsh, and the probability of being taken by the predator is greater for solitary individuals than for individuals in social foraging groups. If individuals have a sufficiently high probability of surviving, calling is more likely to evolve via TFT.

While ornithologists will tell you that many species of birds give food calls to attract others to a newly discovered resource, this behavior has been studied in surprisingly few species. However, because I view food sharing and food calling as prime examples of cooperation, I review what work (theoretical and empirical) exists on this subject. I have intentionally omitted one particular class of examples—namely, those related to the "information center" hypothesis (Ward and Zahavi, 1973). Ward and Zahavi (1973) put forth the information center hypothesis to explain the evolution of colonial nesting in birds. This hypothesis postulates that colonial nesting evolved because individuals could learn about the location of distant food sources from nest mates. Although this hypothesis is fascinating, I have great trouble placing it within the context of cooperative behavior for the following reason: What behavior should we label as cooperative here? That is, given that the information center does not postulate active transfer of information, who is a cooperator and who isn't? Is any bird that returns to the nest a cooperator? What other options are available? Given such questions, I have chosen not to review the information center hypothesis here, and refer the reader to Krebs, 1974; Knight and Knight, 1983; Brown, 1986; Rabenold, 1987; Waltz, 1987; and Gori, 1988, for evidence in support of this hypothesis, and Loman and Tamm, 1980; Pratt, 1980; Kiis and Moller, 1986; Stutchberry, 1988; and Gotmark, 1990, for evidence against it. In some cases, however, I will discuss food calls in species that have been tested in the context of the information center hypothesis, but, again, only when active food calls are given to solicit other group members to forage with a caller at a particular site (e.g., Evans, 1982; Evans and Welham, 1985; Brown et al., 1991).

I also omit two other types of "social foraging" from our discussion; again because although they appear to have superficial cooperative elements, it is difficult to see how one would define cooperators and cheaters in these scenarios. The first case is "beating" (Morse, 1970; Bertram, 1978; Mock, 1980): social foraging in which groupmates feed at high rates because individuals near them cause prey to spring from the ground simply by searching for them. This might be a type of cooperation if "beaters" were attempting to flush insects for others to eat, however the evidence does not support this, but rather that "beating" cannot be avoided by foragers. The second type of social foraging omitted is the case in which increased rates of food consumption per individual are due to *groups* simply foraging at sites with higher prey density, rather than as a result of active hunting by group members (e.g., Krebs, 1974).

4.4.1 Byproduct mutualism in blue jays

One critique of much of the work on reciprocity and the prisoner's dilemma is that although the inequalities of this game are often inferred, and sometimes qualitatively described, they have never been measured precisely (Clements and Stephens, 1995). One imaginative way around such a problem is to construct "Skinnerian" operant psychology-like experiments with precise control over T, R, P, and S. Clements and Stephens (1995) took such an approach when studying foraging in blue jays *(Cyanocitta cristata)*. In their experiments, three pairs of blue jays were tested, and each individual could peck one of two keys—the cooperate key or the defect key. Two payoff matrices were used. The first was a prisoner's dilemma (P matrix), while the second was a "mutualism" matrix (M matrix), in which cooperation was the clear best choice, regardless of what the other player did (Fig. 4.2). Birds were exposed to the P matrix, then to the M matrix, and finally to the P matrix once again. Regardless of whether the jays could see each other or not, a predictable pattern emerged. Birds defected in the first P matrix, cooperated in the M matrix, and reverted to defection the second time they encountered the P matrix.

Blue jays, therefore, cooperated only when $R - T$ and $S - P$ were both positive, and hence their cooperation can best be labeled a case of byproduct mutualism. But why didn't some sort of conditional cooperation occur when they were trapped in a prisoner's dilemma? Clements and Stephens (1995) speculate that it may be because blue jays strongly discount future payoffs, even when they play against each other many times (e.g., in this experiment, during the first exposure to the P matrix, birds played each other an average of 3,435 times). One could, however, argue that jays did not cooperate in this game because it was so removed from how they feed in nature and that natural selection could not possibly have acted on key pecking. While this is always possible, I think that Clements and Stephens (1995) handle this objection nicely, by noting:

> We do not suppose that natural selection has favored key-pecking in Skinner Boxes. Indeed, we offer no argument that blue jays ever faced Prisoner's Dilemmas in their evolutionary history. Rather, our experiment addresses the general significance of the Prisoner's Dilemma as a model of non-kin cooperation. Supporters of the Prisoner's Dilemma have made sweeping claims about its generality (Axelrod and Hamilton, 1981) and have asserted its status as "the leading metaphor for the evolution of cooperation among selfish agents" (Nowak and Sigmund, 1993, p. 56). Given these claims, it is reasonable to suppose that natural selection has equipped animals with the ability to recognize and implement cooperative strategies in novel situations, when economic circumstances favor cooperation. (p. 531)

4.4.2 Food calling in house sparrows

The benefits individuals may gain from locating and joining foraging flocks have been well documented by field and laboratory studies. . . These studies have been concerned primarily with the benefits to individuals that join

M matrix

Player 2

	C	D
C	4	1
D	1	0

Player 1

P matrix

Player 2

	C	D
C	3	0
D	5	1

Player 1

Fig. 4.2. (a) The mutualism payoff matrix (M), and (b) the prisoner's dilemma matrix (P) encountered by foraging blue jays in Clements and Stephens (1995).

> an established foraging flock. However, the process of flock establishment, and the conditions under which individuals may actively attempt to establish foraging flocks have received little attention. (Elgar, 1986, p. 169)

To remedy this, Elgar examined the "chirrup" calls (Summers-Smith, 1963) made by house sparrows (*Passer domesticus*) to test whether this vocalization attracted conspecifics to a food source and, if so, under what conditions. Elgar recorded chirrup calls at artificial feeders which contained a food patch (various bread items). A patch either contained food that was divisible among sparrows, or had just enough food for a single bird. Elgar found evidence that pioneers (those arriving at a patch first) were more likely to give a chirrup call when the food resource was divisible, and that the call rate was higher in these treatments (Fig. 4.3).

Elgar argues that some items were divisible and small enough that sparrows could pick them up and fly away. Larger items, however, were too big to remove from the experimental area. The costs and benefits associated with larger items seem to be such that the anti-predator pluses of calling outweigh the competition minuses of having others at the site. Chirrup calls, then, seem to

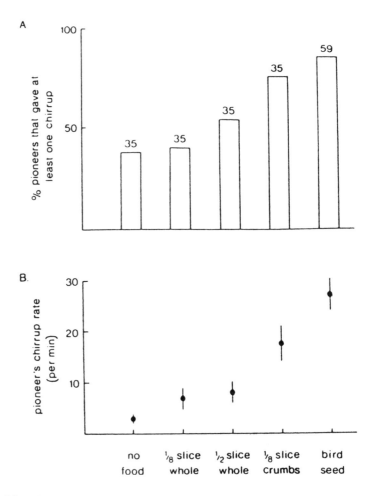

Fig. 4.3. The chirruping behavior of "pioneer" sparrows as a function of food type. Food becomes more divisible as one goes from west to east on the *x* axis. (a) The proportion of pioneers giving at least one chirrup call. (b) The mean chirrup rate of pioneers. Sample sizes are written above columns in A (after Elgar, 1986).

be most readily understood in the context of byproduct mutualism. It is important to note that my argument is not that group foraging per se is explained, but rather that the chirrup call which elicits such formation is best explained by byproduct mutualism.

4.4.3 Raven "yells" and food patches

Heinrich (1988a, b, 1989) and Heinrich and Marzluff (1991) address both proximate and ultimate questions surrounding why common ravens (*Corvus corax*) call when discovering food patches. When discussing such calls Heinrich and Marzluff (1991) caution:

> Whether or not recruitment or information parasitism occurs is often a controversial topic, because to many it raises the question of whether the signallers try to call in others or whether instead they are being exploited. But part of the controversy is artificial because the distinction between proximate and ultimate causes of attraction have commonly not been made. Much confusion has arisen because it is generally assumed that the two coincide. In other words, it is assumed that when animals recruit they give calls "to" attract others. Likewise, if information is parasitized it is usually assumed that the signals are not given "to" attract others. It is not necessary to impart volition to animals giving recruitment signals and then use this as the basis for distinguishing recruitment from parasitism. Attraction of a crowd can yield a great variety of different costs and benefits. But the balance must be positive for attraction signals to evolve. Recruitment is distinguished from parasitism on this purely *functional* ground by investigating the ultimate consequences of group formation. Recruitment results when assembled groups increase the signaler's fitness. Parasitism results when assembled groups decrease the signaler's fitness. (p. 28, emphasis in original)

While I disagree with aspects of this definition, their point about distinguishing ultimate and proximate factors is well taken, but often unexplored. Heinrich (1988a,b, 1989) and Heinrich and Marzluff (1991) illustrate that the proximate reasons animals may call when discovering a food patch may be separate and distinct from the ultimate reason such calls evolve. In ravens, the proximate reason yelling occurs is both a response to hunger level and an advertisement of status (Heinrich and Marzluff, 1991). That is, hungry birds call more often than satiated birds and dominant immatures suppress the calling of those below them. Status may also play a role in the ultimate causes of yelling, as it may translate into increased reproductive success. However, the predominant ultimate causal factor involved in yelling appears to be attracting others to a food resource in order to overpower residents. Most ravens are unrelated "vagrants" (Parker et al., 1994), and the only way a vagrant who comes upon a food source that is being defended can gain access to this source is to "yell," which attracts others, who then together can overpower those originally found at the food source (Heinrich and Marzluff, 1991; see Roell, 1978, for a similar strategy used by jackdaws (*Corvus monedula*) to overpower larger crows at foraging sites).

4.4.4 Active recruitment to foraging sites in cliff swallows

Colonial nesting cliff swallows *(Hirundo pyrrhonota)* provide an excellent example of how to overcome the difficulties of studying cooperation in the context of the information center hypothesis (Ward and Zahavi, 1973). While it is often difficult to delineate which individuals may be "cooperators" or indeed which exact behavior is "cooperative" in the context of this hypothesis, this is not the case for cliff swallows (Brown et al., 1991) and perhaps gulls, *Larus* spp. (Evans, 1982; Evans and Welham, 1985). Using both playback and provisioning experiments, Brown et al. (1991) found that cliff swallows emit a vocalization called a "squeak" call which alerts conspecifics that a food patch has been located. Furthermore, this call appears to be given only in the context of recruiting others to a newly discovered food site.

These squeak calls might best be explained via byproduct mutualism. Cliff swallows feed on ephemeral insect swarms (Brown, 1985) which, when actively grazed, probably do not pose a depletion problem for a caller—that is, callers don't lose food by recruiting others. Those recruited certainly obtain a resource, but also provide the caller with some additional benefits in that with increasing group size it is more likely that one group member will track the insect swarm, and thus provide further foraging opportunities (Brown, 1988; Brown et al., 1991). This tracking behavior may be especially critical, as individual swallows must often return to the colony to care for young, and hence would likely have great trouble relocating an insect swarm without the help of others. Last, the squeak call is given most often during weather conditions associated with poor foraging (the "harsh" environment required for byproduct mutualism; Brown et al., 1991).

4.4.5 Food calls in chickens

> Gallinaceous birds are remarkable and perhaps unique in the animal kingdom for the extent and diversity of the systems of food signalling they have evolved. (Marler et al., 1986a, p. 188)

Male chickens often emit characteristic pulsatile calls when uncovering new food sources (Collias and Joos, 1953; Konishi, 1963; Kruijt, 1964). Marler et al. (1986a,b) and Evans and Marler (1994) examined food calling in male chickens *(Gallus gallus)* to explore the relationship between food calling, "audience effects," and courtship. Marler et al. (1986a) found that the number of calls emitted by a male and the rate at which calls were given vary based on the profitability of food. In addition, females are more likely to approach a male if he is producing a call characteristic of high quality food. Compared with the rate and number of calls emitted when a male is alone, males call significantly more when females are in the area, and significantly less when other males are present (Marler et al., 1986b). Although this suggests that "food calling" may better thought of as courtship behavior, Evans and Marler (1994) found this not to be the case. That is, although males do call more in the presence of females, detailed observations on their courtship behavior indicate that food calling is not part of that repertoire (Evans and Marler, 1994). Despite this, Evans and Marler (1994) suggest that food calling may increase the chance of attracting potentially receptive females (without being part of courtship per se). All of this makes it very difficult to even speculate what category of cooperation is most likely at play in chicken food calls.

4.5 Alarm calls

In principle, alarm calls can be used as a means of deception (Matsuoka, 1980; Munn, 1986; Moller, 1988, 1990) or as a means to manipulate others in a group (Perrins, 1968; Charnov and Krebs, 1975). This, however, may be the (interesting) exception rather than the rule in birds (Klump and Shalter, 1984). Alarm calling, including avian calls, has been considered the quintessential example of altruism, and has received much attention from theorists (e.g., Hamilton, 1964a, b; Maynard Smith, 1965; Trivers, 1971; Charnov and Krebs, 1975; Wilson, 1975; Smith, 1986; Tamachi, 1987; Taylor, et al., 1990). Despite some classic early work on the subject (Marler, 1955), the number of controlled experiments on the function of alarm calling in birds is not as large as one might imagine. In fact, Harvey and Greenwood (1978) note that "No single problem in anti-predator behaviour has attracted more theories and produced fewer facts than that of the evolution and function of alarm calls in bird and mammals" (p. 141).

Things have improved since this statement was made, but not as much as I would have hoped. For example, such questions as, What are the costs and benefits to the caller? To the recipient? Is there any evidence for reciprocity among callers? are often not addressed experimentally in studies on alarm calling. Below I will review some of the available evidence, and how it relates to the evolution of cooperation.

4.5.1. Alarm calls in willow tits

Alatalo and Helle (1990) examined the alarm calls emitted by individual willow tits *(Parus montanus)* exposed to sparrowhawk predators *(Accipiter nisus)*. Willow tits live in small (4 to 8 bird) flocks during the winter (Ekman and Askenmo, 1984; Hogstad, 1987), and group members are not related (Ekman, 1979; Ekman et al., 1981). Alatalo and Helle (1990) set out to examine whether alarm calling in willow tits was dangerous, whether the costs and benefits might differ for group members, and what this might tell us about the role of mutualism and reciprocity in this system.

Willow tits produced alarm calls more often when a hawk was 40 versus 10 m away, suggesting that alarm calls are indeed dangerous to emit (Alatalo and Helle, 1990). But are there differences among individuals in their tendency to produce calls, and if so, why? Alatalo and Helle (1990) speculate that any differences in calling may be tied to the dominance hierarchy found in willow tit flocks (Ekman and Askenmo, 1984; Hogstad, 1987; Koivula and Orwekk, 1988):

Subdominants might actually benefit from the death of a dominant individual, giving them better access to the limited food resources in feeding sites that are safe from predators, or other resources such as roosting cavities, nesting territory and mates. Under these circumstances, if there are no direct benefits of giving alarm calls, such as predator distraction, subdominants might do best by keeping silent, even if there were no cost to calling. Dominants, instead, have more to gain by altering the others, if there are any future benefits. . . . Willow tits usually mate with members of their flock, and because there is a 5–30% surplus of unmated males (Ekman and Askenmo, 1986), the survival of females may be important for males, increasing the benefit to them from calling. (pp. 437–38)

In the language of the cooperator's dilemma, the conditions for byproduct mutualism may be met in dominant, but not subordinate group members.

Alatalo and Helle (1990) found that dominant males were in fact more likely to give alarm calls than other group members, suggesting that they have the most to gain from such behavior. Subordinates do, however, give alarm calls, and it may be that just keeping group size above some critical number may be a great enough benefit to select from some degree of alarm calling in these birds (Smith, 1986). Alatalo and Helle (1990) also suggest that reciprocity per se may play a role in willow tit alarm calls, and while this is tantalizing, no experimental evidence supporting this claim is provided.

4.5.2 Alarm calls and investing in potential mates

In both downy woodpeckers (*Picoides pubescens*) and black-capped chickadees (*Parus atricapillus*), individuals give alarm calls to protect mates, but not other conspecifics. Sullivan (1985) found that downy woodpeckers never gave alarm calls when they were foraging alone (0/46), when in a flock with no other conspecifics (0/23), or when with a same-sexed downy (0/6). In 7 of 9 instances, however, they emitted calls when in the presence of a woodpecker of the opposite sex. Three of these calls were made by females and four by males, indicating the possibility of byproduct mutualism in both sexes.

A similar story can be found in black-capped chickadees in which mated pairs show strong spatial proximity in wintering flocks (Ficken et al., 1980), and alarm calls are probably emitted to increase a mate's probability of surviving the winter (Witkin and Ficken, 1979). In both downy woodpeckers and black-capped chickadees, then, the "harsh" environment that selects for byproduct mutualism is the decreased probability of acquiring a new mate. This "mate investment" has also been suggested to be a causal factor selecting for alarm calls in other birds as well (Leopold, 1977; Morton, 1977; East, 1981; Curio and Regelmann, 1985; Hogstad, 1995). Byproduct mutualism in alarm calling need not be strictly in relation to mates, but could be a more general response to keeping any group members alive (Lima, 1989). For example, Gyger et al. (1986) and Evans et al. (1993) found that chickens *(Gallus domesticus)* were more likely to give a call when in the presence of another individual, but calling was not affected by sex or by pair bonds.

4.6 Mobbing behavior

Mobbing behavior, in which one to a few individuals approach and often chase and/or attack a potential predator, is common in birds (see Curio, 1978; Klump and Shalter, 1984; Sordahl, 1990; McLean and Rhodes, 1991; Dugatkin and Godin, 1992a, for reviews). Curio (1978a,b) has proposed that the primary benefit of mobbing behavior is to force the predator to "move on." A number of other potential benefits are reviewed in Curio et al. (1978a,b), Curio (1978), and Vieth et al. (1980). Considerable evidence, in a number of taxa, supports the move-on hypothesis (Bildstein, 1982; Shedd, 1982, 1983; Buitron, 1983; Donaldson, 1984; Carroll, 1985; Ishihara, 1987; Helfman, 1989; Foster and Ploch, 1990; Pettifor, 1990). Although the costs of mobbing behavior (e.g., how risky is it to the mobbers?) has been the subject of some debate (Hennessy, 1986; Curio and Regelmann, 1986)—similar, I might add, to the debate surrounding the costs of predator inspection in fish—there is considerable anecdotal evidence that mobbing behavior is both quite dangerous (see Sordahl, 1990, for a list of 30 such anecdotes in different species of birds) and quite effective at getting others to join an attack on the predator (Frankenberg, 1981).

4.6.1 Cultural transmission and mobbing in European blackbirds

Given that mobbing appears to have real costs associated with it (Sordahl, 1990), individual group members should likely prefer to be in mobbing groups that are as large as possible. An interesting question that then arises is, What might individuals do to facilitate large mobbing groups? One possibility is that individuals who have experienced a dangerous predator can "teach" their groupmates that a particular stimulus is in fact dangerous and needs to be attacked/mobbed, thereby increasing mobbing group size. Curio et al. (1978b) address this possibility in European blackbirds *(Turdus merula)*. In this experiment, an observer either saw another blackbird attack a novel predator (an Australian honeyeater, *Philemon corniculatus*) or was placed in a control trial. Compared with controls, observers were more likely to try and mob the honeyeater when they saw the model do so (Fig. 4. 4).

In some cases, individuals that had been observers were placed in the role of model, in order to test how long the "chain" of cultural transmission in this system really is. Curio et al. (1978b) found that chains could be six links long—that is, individual 1 could pass on information to individual 2, who passes it on to individual 3, and so on, until individual 7 receives the information! This study supports the notion that mobbing teaches naive individuals of the danger associated with new predators, and hence increases the chances of the "teacher" being part of a larger mobbing group the next time this predator approaches. For more on cultural transmission and mobbing, see Vieth et al. (1980) and McLean and Rhodes (1991).

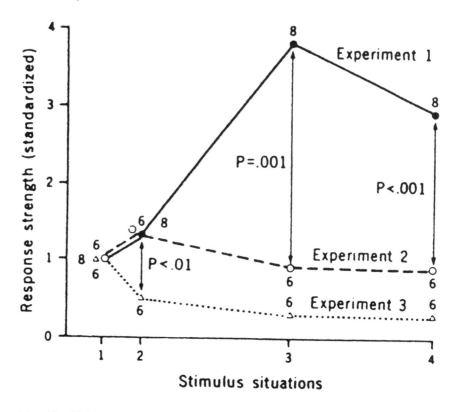

Fig. 4.4. Mobbing response toward honeyeater by observer blackbirds. In experiment 1, the observer saw a conspecific mob a honeyeater. In experiment 2, the honeyeater was shown without a conspecific mobbing it, and in experiment 3, the observer saw an empty box without a conspecific or honeyeater. Stimulus situation 4 measures cultural transmission, and is the data of concern with respect to cooperation (after Curio et al., 1978).

4.6.2 Mobbing, cooperation, and brood defense in great tits

Despite the dangers involved (Curio and Regelmann, 1985), pairs of great tits approach and mob predators in what appears to be a cooperative manner. Regelmann and Curio (1986) begin their examination of mobbing in tits by noting:

> When confronted with a danger pair mates vacillate towards and away from it. They can do so either jointly or alternatively. In joint moves they accrue the benefit of both harassing a predator effectively and deriving the maximum benefit through risk dilution (Hamilton, 1971) and/or predator confusion (Ohguchi, 1981). Moving back and forth alternatively is likely to weaken both these benefits since the predator would be threatened less by only one mate at a time and risk dilution (and probably confusion) is operating with less effect. We are going to examine whether pair members act jointly or alternatively. (pp. 11–12)

Regelmann and Curio (1986) found that 77% of the time, when pairs of great tits were presented with a pygmy owl *(Glacidium perlatum),* both parents mobbed the predator (with males being more likely to mob when only one parent does). When pairs mobbed, it was in a very coordinated fashion, both when approaching the predator and when withdrawing from it (see Bossema and Benus, 1985, for a similar effect in carrion crows, *Corvus corone*). In fact, a significant proportion of an individual's moves were found to be in direct response to the moves of its partner, and individuals responded very quickly to the actions of such a partner during mobbing in tits. Again, given the dangers associated with mobbing in this species, the above clearly suggests that pairs are cooperating with each other. Whether this cooperation is based on byproduct mutualism or reciprocity remains untested, however, as Regelmann and Curio (1986) did not examine the behavior of an individual when its mate did not mob a predator, but it (itself) did (which itself, unfortunately, would be conflated with the differences in mobbing across sexes). Such experiments, designed to examine conditional cooperation, would not, in principle, be difficult to construct.

4.7. Summary

Avian cooperation is all too common. Even if one ignores the large and important literature on cooperative/communal breeding, examples of cooperation are not hard to come by in this taxa. In this chapter I have examined cooperation in the context of territoriality, hunting, food calls, alarm calls, and mobbing, always attempting to tie such examples to current theoretical perspectives on the evolution of cooperation.

5

Cooperation in mammals I: nonprimates

5.1 Introduction

The concept of Scala Naturae (i.e., a natural system of ranking species from high to low) has all but disappeared from the modern-day evolutionary and behavioral literature. Overall, this is probably a good thing. After all, the previous chapters have shown that animals that in the past have been labeled "lower vertebrates" on such a scale are anything but simple. Yet, it is probably still true that *Scala Naturae* lingers in the unconscious, at least with respect to behavior. When pressed, how many behavioral ecologists would not admit that they think that a lion is "smarter" in some appropriate sense than a stickleback? And even though ant colonies are as complex as any nonhuman entity I know of, don't we think that a single ant just isn't as intelligent as a single wild dog?

This book is not about animal intelligence per se, yet the discussion above has implications for the evolution of cooperation. In particular, if mammals are in some sense more behaviorally complex than the groups we have looked at so far, does this not at least suggest that they are cognitively more complex, and doesn't that suggest they are better able to remember events and individuals? If—and this is a big "if"—this string of logic proves to be the case, can't we use it to make some predictions with respect to the distribution of categories of cooperation in mammals and in the groups I have previously touched on? I believe the answer to this question may be "yes," in that reciprocity is the category of cooperation that is most closely associated with cognitive functions such as memory and individual recognition. Although it is true that memory and individual recognition are not always prerequisites for reciprocity (particularly for sessile animals that may pair up for long periods of time), they prove to be critical when interactions between pairs of individuals are interspersed (Axelrod and Hamilton, 1981).

In this chapter, I examine cooperation in nonprimate mammals, reviewing cases of reciprocity, kin selection, group selection, and byproduct mutualism. While the data available may not allow us to quantitatively determine if reciprocity is more common in this group than in fishes, birds, or insects, we can at least (hopefully) get a sense of whether this is true. At the same time, I do not want to get too caught up in this question and overlook all of the fascinating cases of cooperation in the species we are about to examine.

5.2 Grooming behavior

In mammals, grooming behavior appears not only to remove ectoparasites, but to play a role in cementing social relationships. Both these functions may have cooperative elements, and I now review evidence to this effect.

5.2.1 Impala

The strongest evidence of reciprocity in the context of allogrooming comes from Hart and Hart (1992) and Mooring and Hart's (1992) work on impala *(Aepyceros melampus)*. Bouts of reciprocal allogrooming typically occur after one individual begins grooming a nearby conspecific (Fig. 5.1), and grooming does not appear to be solicited. Allogrooming involves the upward sweep of the tongue or the lower incisors along the neck of a partner, and usually involves the initiator performing this activity six to twelve times and the recipient responding in kind. A normal exchange involves three to four such bouts (Fig. 5.2; Hart and Hart, 1992). Grooming appears to reduce the tick load of the individual groomed (see Hart et al., 1992, for more on grooming in African ungulates). The benefit of tick reduction may be very great—if data on the

Fig. 5.1. The start of a sequence of allogrooming exchanges. (Photo courtesy of B. Hart).

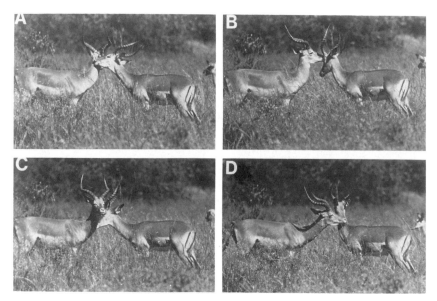

Fig. 5.2. A typical allogrooming sequence showing an exchange between two female impala. Each photo represents a bout of several grooming episodes (after Hart and Hart, 1992).

effects of ticks on the health of cows (Bennett, 1969) is extrapolated to impala, just a few ticks could have serious effects on impala weight and competitive abilities (Hart and Hart, 1992). Grooming itself, however, appears to carry some costs—energy expenditure, electrolyte loss via saliva, and decreases in vigilance (Mooring and Hart, 1995).

The extent of reciprocity within pairs of impala is impressive, to say the least. Whether pairs are male-male, female-female, female-male, subadult male—subadult male or fawn-fawn (Mooring and Hart, 1992), individuals inevitably receive almost the same number of allogrooming bouts that they hand out (Figs. 5.3 and 5.4). Although more controlled experiments on individually recognizable impala are needed to provide concrete evidence that impala are using reciprocity-based strategies, Figures 5.3 and 5.4 certainly hint that something akin to this is being employed during allogrooming. Of course, it is possible that byproduct mutualism may play a role in this system, if ticks provide nutrition or allogrooming builds bonds between individuals (highly unlikely in impala—see below). However, if byproduct mutualism was the driving factor, there is no reason why bouts should end after one partner stops allogrooming, or why both parties in an allogrooming exchange should receive almost the identical number of grooming attempts (but see Connor, 1995b). This, however, is precisely what one would expect if reciprocity was the predominant force in allogrooming. The fact that reciprocal allogrooming is seen in fawns as young as three days old, and that young deprived of allogrooming exchanges with

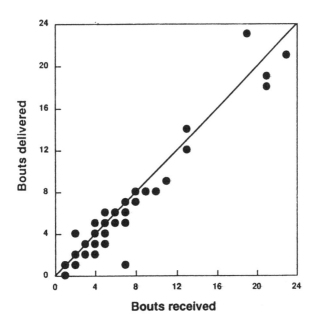

Fig. 5.3. The reciprocal nature of allogrooming among adult female impala. The 45° line denotes a perfect match between bouts delivered and bouts received (after Hart and Hart, 1992).

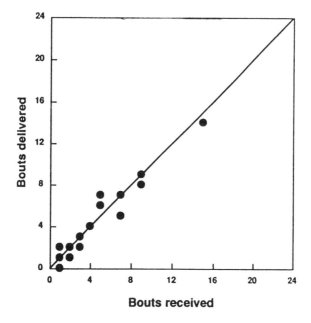

Fig. 5.4. The reciprocal nature of allogrooming among adult male impala. The 45° line denotes a perfect match between bouts delivered and bouts received (after Hart and Hart, 1992).

adults still display this behavior suggests, but is not evidence for, strong selection pressure on reciprocal allogrooming (Mooring and Hart, 1992).

Reciprocal allogrooming in impala has a number of other interesting aspects to it. First, unlike other ruminants, in which allogrooming is typically exchanged among relatives (Clutton-Brock al., 1976; Reinhardt and Reinhardt, 1980), impala undertaking this activity are unlikely to be related (Murray, 1981, 1982), allowing us to rule out a major role for kinship. Second, the iterated aspect of the game here is not the numerous times that individuals are likely to meet (as impala live in very loose social groups; Jarman, 1979), but rather the exchange of many allogrooming activities when pairs come together. Third, unlike the case with primates (Seyfarth, 1980; Fairbanks, 1980) dominant individuals (in male-male interactions) are not more likely to receive a disproportionate number of allogrooming attempts (Hart and Hart, 1992).

5.2.2   Coati

Adult female coati are unique among the procyonid carnivores, in that they form long-lasting social groups (Ewer, 1973; Russell, 1983). Russell (1983) examined two potential types of cooperative behavior in female coati—allogrooming and vigilance—and attempted to distinguish among group selection (which he ruled out immediately because of high immigration and emigration rates), reciprocity, and kin selection as possible mechanisms.

Female coati often groom unrelated individuals within their groups. In order to examine whether reciprocity is involved in such social interactions, Russell examined the average amount of grooming by each individual in a pair and found that it was close to a 50–50 split. While this may hint at some sort of reciprocal grooming (Russell, 1983), such data alone cannot distinguish reciprocity from byproduct mutualism. To do this, one would need to know whether the failure of one party to groom the other changed the nature of the relationship (as it would under reciprocity, but not byproduct mutualism).

5.3   Stotting, tail flagging, alarm signals, and vigilance

As I argued in Chapter 4, alarm signals are an obvious place to look for cooperative behavior. Mammals, however, show an array of different anti-predator signals (see Klump and Shalter, 1984; Caro, 1994), and here I review some of them as possible cases of cooperative behavior.

5.3.1   Stotting in Thomson's gazelles

Stotting behavior, wherein individuals take all four legs off the ground simultaneously and hold them straight and stiff in the air (Caro, 1986b; Walther, 1969), has been documented in a number of species of Cervidae, Antilocapridae, and Bovidae (Byers, 1984). While typically occurring in the presence of a predator, stotting may also take place in the context of play behavior (Byers, 1984). The most thorough studies of the costs and benefits of stotting are Caro's (1986a,b) work on Thomson's gazelles *(Gazella thomsoni).*

Caro (1986a) examined whether any time, energy, and/or survivorship costs were associated with stotting in Thomson's gazelles and tested which of eleven nonmutually exclusive hypotheses best explains the function of stotting (Caro, 1986b). Somewhat surprisingly, he found that although a survivorship cost was historically associated with stotting (Estes and Goddard, 1967), gazelles stot from "safe" distances (outside the attack range of potential predators), and no stotting gazelle was caught by a cheetah during Caro's relatively long-term study. Of the eleven hypotheses Caro (1986a) examined, only two received strong support. Stotting seemed to inform predators that they have been seen, and stotting in juveniles attracted their mother's attention in dangerous situations. Subsequent work by FitzGibbon and Fanshawe (1988) has provided some support for the notion that stotting informs a predator that the stotter is in good health (as predicted by Zahavi, 1975; also see FitzGibbon, 1994, for more on the costs and benefits of predator inspection in gazelles).

It appears then that stotting may not be cooperative behavior at all, but rather a selfish strategy that increases the stotter's chance of survival and which has *little or no effect on other group members.* That is, although no data is available on whether groups with stotters have higher survival rates than groups without stotters, it appears that potential predators "lock" onto an individual at the start of a hunt, and the hunt succeeds or fails dependent on whether that individual is taken (Caro, personnel communication).

5.3.2 Tail flagging and snorting in white-tailed deer and other ungulates

In many species of artiodactyle and lagomorph ungulates, individuals are known to "flag" their tails after a predator has been sighted. Such flagging behavior occurs as a part of a sequence of anti-predator behaviors, and often involves an individual lifting its tail and "flashing" a conspicuous white rump patch. Flagging often, but not always, occurs when a predator is at a relatively safe distance from its potential prey (Hirth and McCullough, 1977). As with stotting behavior, a number of alternative hypotheses have been put forth to explain the functional significance of tail flagging (see Harvey and Greenwood, 1978, for more).

Tail flagging has been postulated to: (1) warn conspecifics (kin and nonkin) of potential dangers (Estes and Goddard, 1967); (2) "close ranks" and tighten group cohesion (McCullough, 1969; Kitchen, 1972; Smith, 1991), perhaps to ensure group-related foraging and anti-predator benefits in the future (Trivers, 1971; Smith, 1986); (3) announce to the predator that it has been sighted and should therefore abandon any attack (Woodland et al., 1980; Caro, 1995a; Caro et al., 1995); (4) entice the predator to attack from a distance that is likely to result in an aborted attempt (Smythe, 1977); (5) cause other group members to respond, thereby confusing the predator and making the flagger itself less likely to be the victim of an attack (Charnov and Krebs, 1975; this hypothesis was generated for the general case of alarm calls, but can be applied to tail flagging); or (6) serve as a sign for appeasing dominants, and play only a secondary role in anti-predator behavior (Guthrie, 1971). Some of these hypotheses (and their variants) present tail flagging as a cooperative behavior, while others do not.

Data from studies on white-tailed deer *(Odocoileus virginianus)* unfortunately do not fit squarely into any of the above hypotheses. Hirth and McCullough (1977) argue that doe-doe groups have greater coefficients of relatedness than do buck-buck groups, and that inclusive fitness theory would predict more tail flagging in the former than the latter. This proved not be to the case with tail flagging, but was true for snorting, another alarm signal that is often associated with the presence of a predator. Why kinship should affect one behavior that alerts group members of a potential danger but not another remains unclear. Aside from kinship effects, Hirth and McCullough (1977) and Smith (1991) found support for the hypothesis that tail flagging causes an increase in group cohesion and yields group benefits associated with such behavior. This finding might be construed in the context of byproduct mutualism. That is, group members flag to keep group size large, thereby benefiting themselves, as well as others, in the group. It is important to note this is not an argument put forth to suggest that byproduct mutualism explains the evolution of grouping behavior in general, but rather that when behaviors evolve that increase the likelihood of continued group membership, byproduct mutualism may be playing a role.

The story with respect to flagging in white-tailed deer is, however, even more complex. Bildstein (1983) could not rule out the cohesion hypothesis, but argued that tail flagging was directed at the predator, not at others in the group, and was likely a signal to the would-be predator that an attack was not likely to succeed. Tilson and Norton (1981) put forth a similar argument for the function of alarm duetting in pairs of klipspringers *(Oreotragus oreotragus)*. This pursuit-deterrence argument (Woodland et al., 1980) is supported by Caro et al.'s 1995 work on white-tailed deer. Caro et al. found no evidence for a time cost to flagging (also see LaGory, 1987), nor was tail flagging aimed at conspecifics (in the context of cohesion or manipulation), including kin. Rather, it appears that individuals that run fast flick their tails and are using this signal to communicate to the predator that an attack is unlikely to succeed.

Without data on the success rates of predators on flaggers and nonflaggers, it is difficult to evaluate which, if any, mechanisms of cooperation might account for signaling to a predator. If such signals increase the survivorship of the signaler alone, then tail flagging is not a cooperative act; the same argument can be made for barking alarm calls in barking deer *(Muntiacus reevesi;* Yahner, 1980). However, if this signal causes a decrease in attack rates on all group members, as appears to be the case in Tilson and Norton's (1981) work on pairs of klipspringers, then group selection, byproduct mutualism, or kin selection are still respectable candidates for explaining the evolution of tail flagging.

5.3.3 Alarm calls in squirrels

One of the most well-known and well-cited examples of kinship-based cooperation is Sherman's (1977, 1980, 1981, 1985) work on alarm calls in Belding's ground squirrels *(Spermophilus beldingi;* Fig. 5.5). Studying an individually

Fig. 5.5. A marked female Belding's ground squirrel emitting a predator alarm call. (Photo courtesy of P. Sherman).

marked population of Belding's ground squirrels, in which most pedigrees were well known, Sherman (1977) addressed which of six hypotheses best explained the evolution of alarm calling in this species. Although he addresses each of the four mechanisms that can lead to cooperation, in retrospect, his hypothesis 3—*alerting relatives*—turns out to be the most interesting. Sherman hypothesizes:

> Callers may gain by having placed themselves in some jeopardy if kin are thereby consistently warned. Captured individuals may also give distress (alarm) calls in this context, thereby soliciting assistance from relatives, or else warning them to flee or hide. Under this, the third hypothesis, year-round alarm calls must be associated with the continuous presence of relatives. . . . If alarm calls are given during only part of the year, they must coincide with the proximity of kin. For a given species, this hypothesis would be strongly supported if individuals with relatives living within earshot call more frequently than do conspecifics without them. (p. 1246)

In support of this hypothesis, Sherman found that female squirrels gave alarm calls when a predator was in the vicinity more often than expected by chance, while the converse was true for males (Fig. 5.6). Females are generally sedentary (with respect to emigration) and mature and breed near their natal

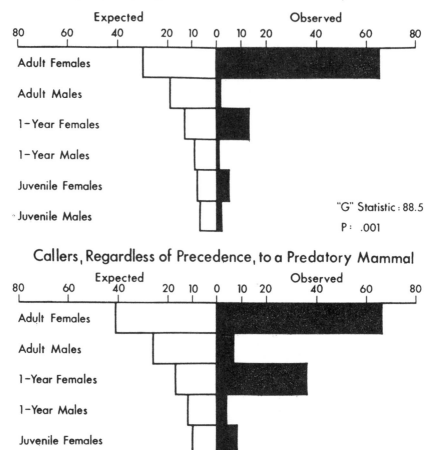

Fig. 5.6. Expected and observed frequencies of alarm calling in Belding's ground squirrel. The overall significance of both comparisons is due to females calling more than expected and males calling less. Data based on 102 observations (after Sherman, 1977).

sites, while males always emigrate from their birthplace and don't aggregate with siblings after emigration (Sherman, 1977). As such, females were warning close kin (often offspring) by giving such alarm calls, while no such benefit accrued to males for warning others about the presence of a potential predator. Further support for the kinship hypothesis includes evidence that "invading" (non-native) females gave alarm calls less frequently than native females. This evidence also demonstrates that females with living relatives are more likely to call than females without any living kin (Sherman, 1977).

Kinship also appears to play a role in fighting and chasing in ground squirrels (Sherman, 1981), but does not affect alarm calling in response to aerial predators (Sherman, 1980, 1985). Other examples of kin-selected alarm calling in rodents include Barash (1975, 1976), Dunford (1977), Leger and Owings (1978), Schwagmeyer (1980), Smith (1978), Hoogland (1983, 1995), and Davis (1984).

Kinship, however, need not be the only category of cooperation associated with alarm calling in mammals. Female coati group members give off alarm calls when a predator is nearby (Janzen, 1970). Since coati groups are often made of related individuals, Russell (1983) addressed the question of whether the degree to which group members were related affected the frequency at which alarm calls were given. Using a very small sample set (two bands), he compared the degree of alarm calling in a band in which relatedness among band members was high with one in which relatedness was low. No differences were found, indicating either that kinship per se is not a critical parameter in the decision to give an alarm call or that individuals are not able to distinguish between kin and nonkin.

5.3.4 Circular formations and vigilance in elephants

The popular depiction of the "friendly" elephant may not be that far from the truth, at least for female African elephants *(Loxodonta africana)*. Although reproductive competition is clearly a factor in elephant groups (Dublin, 1983), cooperation is more the rule (Douglas-Hamilton and Douglas-Hamilton, 1975; Dublin, 1983). Female elephants typically remain in maternal family units and such female units are complex societies led by a matriarch (usually the largest elder female; Moss, 1976).

Dublin (1983) and Lee (1987) present evidence for cooperation in female African elephants in the context of vigilance behavior. When not in circular defense formation (with juveniles in the center), the matriarch and larger individuals will take a lead position (or guard the rear depending on the situation) when a confrontation with potential danger occurs (Douglas-Hamilton and Douglas-Hamilton, 1975). Dublin (1983) notes that since groups of elephants contain siblings and mothers and daughters, kinship surely plays a role in such cooperation. She also notes that reciprocity may play a role in cooperative group vigilance, but provides no empirical evidence that this is the case. Rather, Dublin argues that since female elephants are often in the same group for all or at least most of their lives, the "shadow of the future" *(w)* is long and consequently favors the evolution of cooperation via reciprocity.

5.4 Coalitions

5.4.1 Byproduct mutualism and kinship among male lions

Coalitions of male lions *(Panthera leo)* cooperate by patrolling the pride's territory and chasing off other male intruders (Packer et al., 1991). Individuals in larger coalitions father more offspring per male pride member (Packer et al., 1988), and early work on coalitions and cooperation among male lions focused on byproduct mutualism-like explanations for such alliances. That is, early work assumed that all males sired an equal proportion of offspring, and that each male consequently gained fitness benefits when group size increased (Bygott et al., 1979; Packer and Pusey, 1982; Packer et al., 1988). Packer et al. (1991), however, found that byproduct mutualism appears to be important only in *pairs* of male lions that join together. Such pairs are often composed of unrelated individuals, each of which fathers offspring in their pride. In larger coalitions, however, kinship is the primary force selecting for cooperation (Packer et al., 1991). Given that larger coalitions produce more offspring per male, why don't unrelated males form larger coalitions? The answer appears to be that in larger coalitions, the variance in reproductive success increases, with a single male often siring all the offspring (Packer et al., 1991; Fig. 5.7). As such, indirect kinship benefits seem to allow larger coalitions among relatives, but not among nonkin (Packer et al., 1991).

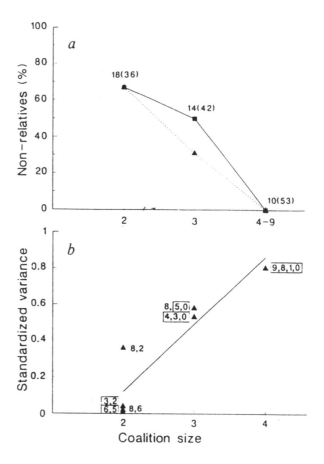

Fig. 5.7. (a) Percentage of coalitions containing nonrelatives (darkened squares) and of males lacking a related partner, as function of coalition size. (b) Standardized variance in reproductive success as a function of group size (after Packer et al., 1991).

5.4.2 Alliances and "herding" behavior in cetaceans

Pairs and triplets of males form close associations within the fission-fusion society of the bottle-nosed dolphins (*Tursiops truncatus*) of Shark Bay, Western Australia (Connor et al., 1992a,b). Connor et al. (1992b) found two levels of "alliances" between males "herding" reproductive females. Alliances involve males acting in a coordinated fashion to keep females by their side, presumably to mate with them. Should females attempt to escape:

> The males chased in 25 (56%) of the escapes. The manner in which males chased bolting females illustrates the cooperative nature of herding: rather than chasing directly behind the female, the males often angled off to either side, effectively cutting the distance if she changed direction. (p. 987)

First-order alliances included pairs and triplets of males that consistently acted as a unit to herd. Furthermore, members of a first-order alliance associated closely when not in the act of herding females. Second-order alliances were composed of pairs of first-order alliances that joined together and aggressively attacked and "stole" females from other alliances. Which first-order alliances joined together apparently depended on a number of unknown factors, and was not consistent over time (Connor et al., 1992b). Until more is known about the degree of relatedness among alliance members, and the distribution of matings within alliances, little more can be said on which category of cooperation, if any, best explains alliance formation in dolphins.

5.5 Alloparenting and helping behavior

Alloparenting—the dispensing of "parental" behavior to young that are not one's own—is quite common in nonprimate mammals. Perhaps the most well-studied type of alloparenting is "helping" behavior. Helping is typically undertaken by young, related individuals, who are physiologically capable of reproduction but, for various reasons, refrain from it (see Chapter 4 and below for references). Other varieties of alloparenting are also widespread in nonprimate mammals. Many of these cases involve adults that are either reproductively active or have been in the past, and are often unrelated to the young they care for.

In mammals, as in birds, a wide variety of nonmutually exclusive mechanisms have been put forth to explain the evolution of helping behavior. Following Emlen (1991), Jennions and MacDonald (1994) provide a list of nine hypotheses for the evolution of helping behavior, with suitable mammalian examples for each. In Table 5.1 I have modified Jennions and MacDonald's (1994) list, and in so doing have attempted to examine how each of these hypotheses fits into our four categories of cooperation (reciprocity, group-selected cooperation, kin selection, and byproduct mutualism).

Our discussion of helping in mammals begins with the following caveat:

> In mammals, little field-work, but much theorizing, has investigated the effect of variation in territory quality on individual dispersal decisions and

Table 5.1. Hypotheses for the evolution of helping in mammals. In the last column, I have speculated which category of cooperation might best fit the example. BPM = byproduct mutualism, GS = group selected cooperation, KS = kin selection, and RA = reciprocal altruism (after Table 3 of Jennions and McDonald, 1994).

Benefit to helper	Proposed mechanism	Mammalian examples	References	Categories
Survivorship	Group size improves vigilance, anti-predator behavior, or feeding success.	In evening bats, females allosuckle unrelated pups (predominantly females), increasing colony size and future access to information about feeding.	Wilkinson, 1992	BPM
	Helping is payment for access to the natal territory.	In the naked mole rat (*Heterocephalus glaber*) there is conflict between the queen and the workers over the amount of work performed, suggesting that some payment is extracted (although conflict may occur for other reasons as well).	Moehlman, 1989 Reeve, 1992	BPM, KS
Future probability of territory holding	Larger groups expand their territories.	Naked mole rat colonies probably divide by fission. The likelihood that the daughter colony will succeed is probably related to the size of the work force.	Sherman et al., 1991	GS
Future probability of breeding	Helping results in coalitions between donors and recipients, and coalitions are more likely to obtain vacant territories than are individuals.	In dwarf mongooses (*Helogale parvula*), recipients and donors of help may disperse together. In lions (*Panthera leo*), mothers increase the likelihood of their offspring obtaining coalition partners by allosuckling. Larger coalitions are more successful (but also tend to comprise closely related males due to the increased skew in mating success).	Packer et al., 1988; Rood, 1990; Packer et al., 1991	GS, KS

Reproductive success	Helpers gain breeding experience, which increases their own breeding success.	In many mammals, more experienced mammals have increased success in raising young. However, it remains to be shown that helpers are more successful than nonhelpers when they first breed.	Rasa, 1987	BPM
Reproductive success	Helpers are more likely to gain the support of recipients of help as future breeders.	In dwarf mongooses, many helpers eventually breed in their natal group. Given that delayed dispersal is common, helpers will thus gain a future direct benefit. However, it remains to be shown that this benefit is greater than the initial cost of helping, which requires a comparison between helpers and nonhelpers.	Rood, 1990; Creel and Wasser, 1991	RA
Production of nondescendant kin	Increased survival of related breeders, hence higher reproductive success for breeders.	Breeders have higher survivorship when helpers are present in dwarf mongooses and lions.	Packer et al., 1988; Rood, 1990; Packer et al., 1991	KS
	Increased survival of recipients of help which are related to the helper.	In most cooperative breeding mammals, helpers and donors are closely related, and substantial indirect benefits may be realized.	Creel et al., 1991; Packer et al., 1991	KS

experimental manipulation of territory quality is likely to be a profitable
line of future research. (Jennions and MacDonald, 1994, p. 90)

As such I concentrate here on why individuals help, given that they have
the opportunity to do so, rather than whether they stay on natal territories because of limited alternatives.

As Creel and Creel (1991) note, helping in mammals may take the form of direct benefits to young as in, for example, blackbearded jackals (*Canis mesomelas;*
Moehlman, 1979), wild dogs (*Lycaon pictus;* Malcolm and Marten, 1982), dwarf
mongooses (*Helogale parvula;* Rood, 1978), lions (*P. leo;* Schaller, 1972), and
brown hyena (*Hyaena brunnea;* Owens and Owens, 1984), or direct benefits to
the parents as in, for example, silverback jackals (Moehlman, 1989).

5.5.1 Kinship, reciprocity, and helping in brown hyenas

Brown hyena *(H. brunnea)* "clans" in the Kalahari Desert typically contain
numerous adult and subadult males and females, yet a single female will typically bear all the cubs in a given breeding season (Owens and Owens, 1984).
Females move cubs from "minor dens" to communal dens when the cubs are
about 2.5 to 3 months old (Owens and Owens, 1979, 1984). In such communal
dens, 79% of the provisioning is done by aunts, cousins, and half-sibs (Owens
and Owens, 1984). Most of this help comes from females, who are willing to
provision kin that were more distantly related than the recipients of male help.
Immigrant males (who sire most offspring in a clan) were never seen provisioning young.

While provisioning clearly provides indirect fitness benefits via kinship, reciprocity may also play a role in the evolution of cooperation in this system.
Owens and Owens (1984) note: "Females who received help when they were
cubs later provisioned their helper's offspring and adult females of similar ages
provisioned one another's cubs" (p. 844).

Although there is no evidence that any safeguards exist in this intergenera-
tional exchange system, it is an interesting possible example of reciprocity
nonetheless (see Smale et al., 1995, for more on cooperation and aggression
among spotted hyena *(Crocuta crocuta)* litter mates).

5.5.2 Kinship, byproduct mutualism, and helping in
naked mole rats

Few mammals have captured the fancy of both scientists and layman to the
extent of the naked mole rat *(Hetercephelus glaber)*—the first eusocial vertebrate ever uncovered (Jarvis, 1981; Sherman et al., 1991). These small hairless
rodents of tropical Africa were the first vertebrate to meet the strict criteria that
define eusociality—a reproductive division of labor, overlapping generations,
and communal care of young (Fig. 5.8). Naked mole rats live in large groups
(Fig. 5.9) in which a single queen and 1 to 3 male reproductives are in possession of exclusive breeding rights for the colony, and while intra-colony aggression (manifest in queen-worker conflict) certainly exists (Reeve and Sherman,
1991; Reeve, 1992), cooperation is the order of the day.

Fig. 5.8. Frontal view of a naked mole. Individuals are hairless and use the sharp teeth seen in this photo to excavate a labyrinth of underground tunnels. (Photo courtesy of S. Braude)

Fig. 5.9. Part of a colony of naked mole rats (viewed from above). The workers shown in this photo are all highly related and are considerably smaller than the one reproductive queen/colony. (Photo courtesy of S. Braude).

Nonreproductive males and females, who in fact live on average much shorter lives than reproductives (Sherman et al., 1992), are involved in various sorts of helping behaviors, such as digging new tunnels for the colony (a critical aspect of colony survival), sweeping debris, grooming, and predator defense (see Lacey et al., 1991; Lacey and Sherman, 1991; Pepper et al., 1991, for ethograms and details on division of labor in naked mole rats). Yet why should individuals in a respectable diploid species yield exclusive reproduction to a single queen and a few males? This question itself may be misleading, as nonreproductives do not so much yield reproduction as they are coerced into this via aggression on the part of the queen (Faulkes et al., 1989, 1990, 1991). This, however, raises the issue of why nonreproductives have not developed better mechanisms for overcoming queen coercion. Until recently, the answer appeared to be connected to kinship per se. That is, even though naked mole rats are diploid, average relatedness in a colony is $r = 0.81$ (higher than the average r among hymenopteran sisters!; Reeve et al., 1990: also see O'Riain et al., 1996 for evidence of a "dispersive morph" in mole-rats). In this system, then, the indirect benefits of kinship were thought to underlie the cooperative nature of nonreproductives. It turns out, however, that a comparison of naked mole rats with Damarland mole rats *(Cryptomys damarensis),* who are also eusocial, shows that the indirect benefits of kinship may only be part of the picture, and that byproduct mutualism may also play a role in this system. Jarvis et al. (1994) note that ecological factors, particularly rainfall, have played a large role in the evolution of cooperation in naked mole rats and in Damarland mole rats:

> Throughout the evolutionary history of the Bathyergidae, successive prolonged droughts may have thus resulted in "ecological bottlenecks" which served to exclude the solitary genera from arid regions. Only those species in which the young did not disperse at weaning, but rather remained at home to help were able to survive in such an unpredictable environment. Some of the seemingly bizarre features of naked mole rats—for example, lack of hair, poikiothermy, extreme inbreeding and huge colonies—may be red herrings to our understanding of their social evolution. After all, Damarland mole-rats share none of these attributes but are also eusocial. The unique features of naked mole-rats probably became possible after their highly social lifestyle evolved, and because of the buffered underground niche they occupy. (p. 51)

Couched in the language of our theoretical construct, one might say that at the evolutionary onset of cooperation in naked mole rats, when reproductive division of labor was likely minimal, a "harsh environment" central to byproduct mutualism, rather than kinship per se, may have been the predominant selective agent. It is possible that only after communal life became obligate, and inbreeding increased, that kinship magnified cooperative behavior in naked mole rats.

5.5.3 Kinship, pseudoreciprcoity, reciprocity, and helping in dwarf mongooses

Dwarf mongooses (*H. parvula*) of the Serengeti National Park live in groups with a median size of 9.5, and the alpha male and female are usually (but not always) the parents of all group members born during a given breeding season (Rasa, 1977, 1987; Rood, 1978, 1983). In *H. parvula* helping takes numerous forms—feeding, defense, grooming, and transporting the young (Fig. 5.10)— but the most interesting variant of helping in this species may be "baby-sitting" (Rood, 1978; also see Rood, 1974, for more on baby-sitting males in the banded mongoose *(Mungos mungos)* and in meerkats *(Suricata suricata))*. When mongoose packs go out and forage, a "baby-sitter" stays back and watches the newborn young. Who plays the role of baby-sitter switches often, but it is a rare occurrence for no baby-sitter to be present when young are in the den. Often sexually mature, but not reproductively active, yearling females serve as baby-sitter. This is probably due to baby-sitting being a particularly effective anti-predator technique, as baby-sitters have been seen giving alarm calls and even chasing potential predators from the den (Rood, 1978).

What benefits do helpers obtain by helping? Since all young in a den are usually related, helpers are no doubt receiving some indirect kinship benefits from their actions. Kinship, however, does not explain the whole picture. Rood (1978) found that when a two-year-old immigrant female, who was presumably

Fig. 5.10. A group of 7 adult and 2 juvenile dwarf mongooses in the wild. The pair of adults in the upper right corner are engaged in agonistic behavior. The alpha pair at the bottom are engaged in sexual behavior, while the adult and juvenile in the center are involved in typically "friendly" interactions. (Photo courtesy of S. Creel.)

unrelated to any pack members joined a group, she was the most active baby-sitter of the lot. Helping, however, may have elements of pseudo-reciprocity (Connor, 1986) and byproduct mutualism to it, as it increases future group size (Rood, 1990; Fig. 5.11), and the recipients of help today are the alarm callers of tomorrow. That is, helping increases group size, which in the future when helpers mature, provides the "many eyes" needed to detect predators (Pulliam, 1973).

"True" reciprocity may also play a role in the evolution of helping in dwarf mongooses. Rood (1983, 1986) suggests that helpers may be reciprocated for their help in three different ways when those they helped mature. Former recipients of help may: (1) assist helpers in defending their territories, (2) form an alliance with the former helper in order to acquire a new territory, and (3) help to raise the young of the former helper (see Ligon, 1991, for a comparison of cooperation in dwarf mongooses and green wood hoopoe).

5.5.4 Allomothering in elephants

Nulliparous female elephants devote large chunks of time to allomothering (e.g., guarding the young of other females; Douglas-Hamilton and Douglas-Hamilton, 1975; Leuthold, 1977; Kingdon, 1979; Lee, 1987). Dublin (1983) offers a variety of possible reasons for such cooperative behavior. The most obvious of these is kinship, in that females are likely caring for maternal or paternal half-siblings, and accruing indirect fitness benefits from increasing their siblings' probability of survival above and beyond what it would be without allogrooming. Interestingly, allomothering appears to increase when times are hard (illness, injury, etc.), suggesting that this behavior is dispensed to kin when it is most needed (unpublished data cited in Dublin, 1983).

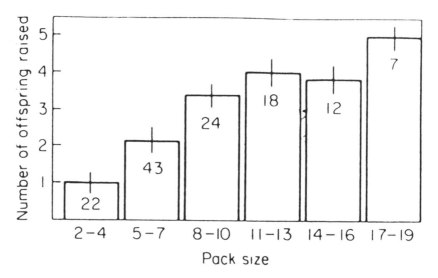

Fig. 5.11. Number of offspring raised to yearlings as a function of group size in dwarf mongooses (after Rood, 1990).

In addition to kinship, other factors probably favor the evolution of allo-mothering as well. While no direct evidence supports it, females may be gaining experience at raising young—a difficult and time-consuming job in elephants. Given the disastrous effects of losing an offspring that gestates for 22 months, experience at raising young is no doubt important. But perhaps the clearest benefit of allomothering is that subdominate females that undertake such behavior are rewarded by dominant females with access to valuable resources. Given the importance of such resources in an environment that is very patchy, and the difficulty of moving from herd to herd, byproduct mutualism likely plays a role in selecting for allomothering among group members.

5.5.5 Bats, midwives, and cooperation

Kunz and Allgaier (1994) describe a fascinating case of allomaternal care in the Rodriques fruit bat *(Pteropus radricencis)*. In this species, unrelated females assist pregnant individuals in the birthing process (Fig. 5.12). Although sample size in this study was small, Kunz and Allgaier (1994) found that those assisting in the birth were involved in licking (of the pregnant individual) and in wrapping their wings around expectant mothers. In addition, mothers-to-be didn't go into the typical feet-down birthing position until helpers took this position. While the data do not allow us to determine how cooperation evolved in this system, the authors suggest that because *P. radricencis* are both long-lived and extremely social, reciprocity may very well play a role in assisted births.

Kunz and Allgaier (1994) note that anecdotal evidence for assistance during birth has been recorded in marmosets (*Callithrix jacchus;* Lucas et al., 1937), Indian elephants (*Elephas maximus;* Poppleton, 1957), African hunting dogs (*Cunis familiaris;* Gaffrey, 1957), Japanese dogs (*Nyctereutes procyonoides;* Yamamoto, 1987), spiny mice (*Acomys cahirinus*), and bottle-nosed dolphins (*Tursiops truncatus;* Essapian, 1963).

Another fascinating example of allomothering in bats is the case of communal nursing in the evening bat *(Nycticeius humeralis)*. Although most nursing bouts in this species involve mother and pup, and even though females will reject pups when they do not wish to nurse, Wilkinson (1992a) observed that approximately 20% of nursing bouts involved females feeding unrelated pups. Why should nursing unrelated pups be so common in this species? Wilkinson (1992a) suggests that females nursing unrelated pups may be receiving immediate, as well as delayed, benefits for their actions. One immediate benefit to such a female is that if her pups have already fed to satiation, nursing unrelated pups may decrease her weight during foraging bouts in the immediate future. In addition to being able to forage more efficiently when such "extra" milk is disposed of, storing milk in mammary glands may suppress future milk production (Mepham, 1976) and increase the chances of infection. Milk "dumping" then may provide another substantial immediate benefit to communal nursing.

Even nursing of *unrelated* pups in *N. humeralis* is, however, not unselective. Females are more likely to nurse unrelated female pups than unrelated males, and this may be due to the fact that males of this species disperse, while females do not (Wilkinson, 1992a). This demographic parameter may be im-

Fig. 5.12. A pregnant female Rodriques fruit bat ready to give birth. (Photo courtesy of T. Kinz.)

portant in that information transfer during foraging bouts increases foraging success in *N. humeralis* (Wilkinson, 1992b). By nursing unrelated female pups, females are helping those individuals who are likely to provide useful foraging-related information in the future.

Given the enormous costs of lactation in many mammals (Millar, 1977; Gittleman and Thompson, 1988), it is surprising to note the frequency of allosuckling behavior (suckling non-descendant kin) in this taxa. For a comparative analysis of (nonoffspring) nursing in 100 species of mammals, see Packer et al. (1992); for other examples of allomaternal behavior in general, the reader is directed to Reidman's (1982) review on adoption and allomaternal care in more than 100 species of mammals, as well as Gittleman (1985).

5.5.6 Allomothering in dolphins?

Tavolga and Essapian (1957) observed allomaternal care in the birthing process of the bottle-nosed dolphin *Tursiops truncatus*. Before parturition, bottle-noses form stable associations with another adult female (the "auntie dolphin"; Norris and Schilt, 1988). This other female provides assistance during birth itself and the early rearing period. The auntie dolphin may even be the individual that brings the newborn to the water's surface for its first breath (Tavolga and Essapian, 1957).

5.5.7 Other examples of helping and alloparenting

Again, as was the case with birds, I cannot possibly discuss every example of cooperation in the context of mammalian helping, but rather have chosen representative examples and direct the reader to MacDonald and Moehlman (1982), Reidman (1982), Emlen (1984), Gittleman (1985), Creel and Creel (1991), Packer et al. (1992), Jennions and MacDonald (1994), and Solomon and French (1996) for more comprehensive reviews on helping in mammals.

5.6 Cooperative hunting

Cooperative hunting is extremely common in mammals, common enough, in fact, to merit a book of its own (see Packer and Ruttan, 1988, and Caro, 1995b, for more extensive reviews). While cooperative hunting is well documented in many mammals, only in the last five years have these studies been designed to test any models for the evolution of cooperation. The most comprehensive tests of Packer and Ruttan's (1988) models for the evolution of cooperation (see Chapter 2) have been undertaken using lions (*P. leo;* Fig. 5.13), and I concentrate on this work here (see Schaller 1972; Caraco and Wolf, 1975; Clark, 1987, for early work on group size in foraging lions).

Scheel and Packer (1991) note that until recently, many behavioral ecologists believed that hunting in lions was a quintessential case of cooperation. Yet

Fig. 5.13. A pair of lionesses guard a buffalo they have just successfully hunted. (Photo provide by Anthro-Photo.)

detailed observations of hunting in this species indicate that some pridemates are clearly exploiting their cooperating compatriots (Scheel and Packer, 1991; Heinsohn and Packer, 1995). Scheel and Packer (1991) examined the hunting behavior of lions in the Serengeti National Park and, as will become obvious below, found that cooperative hunting in lions is best understood in terms of byproduct mutualism (also see Grinell et al., 1995).

Studying the hunting behavior of primarily female lions, Scheel and Packer (1991) tested four hypotheses generated by Packer and Ruttan's (1988) game theory models of hunting behavior:

(1)   The tendency to participate in hunts will increase with the difficulty associated with hunting a particular prey.

(2)   Poor hunters should hunt less. This is predicted because the benefits of larger hunting groups increase less steeply with the addition of poor hunters; hence, these benefits may not outweigh the costs of joining a hunt for poor hunters. Although not the subject of Scheel and Packer's (1991) study, in the same light, hunting groups may actively try and stop poor hunters from joining in a hunt.

(3)   Because of the asymptotic relationship between per hunter benefit and group size, individuals should be less likely to join large hunting groups.

(4)   Cooperation should be more likely among kin than nonkin.

Although prediction 4 clearly relates to kinship, the first three seem to be generated via the logic of byproduct mutualism. It turns out that Scheel and Packer's data on hunts of various different prey types support predictions 1, 2, and 3, but not 4, lending an even greater air of byproduct mutualism to lion hunting behavior. Scheel and Packer did more, however, than simply test the above predictions. One of their principle goals was to test whether some lions were more likely to cheat on hunts than others. Using finite mixture models, which essentially try to determine which combination of underlying statistical distributions best explains a data set, Scheel and Packer found that three strategies appear to be associated with hunting in lions—refraining (not hunting), conforming (all hunters behave similarly), or pursuing (actively hunting) (Fig. 5.14). Whether refraining equals cheating, conforming equals conditional cooperation, and pursuing equals pure cooperation remains to be seen, although Packer and Ruttan (1988) found no evidence for TFT in the context of cooperative hunting, making the second equality unlikely.

Heinsohn and Packer (1995) have recently shown that cooperative territory defense in female lions is even more complex than described above. They found evidence that some individuals lagged behind others when approaching dangerous intruders, and defined four strategies; (1) unconditional cooperators, who always lead when approaching a potential danger, (2) unconditional laggards, who consistently follow leaders, (3) conditional cooperators, who cooperate only when they are most needed, and (4) conditional laggards, who defect when they are most needed. Heinsohn and Packer (1995) argue that these strategies in conjunction with other evidence on territory defense in female lions

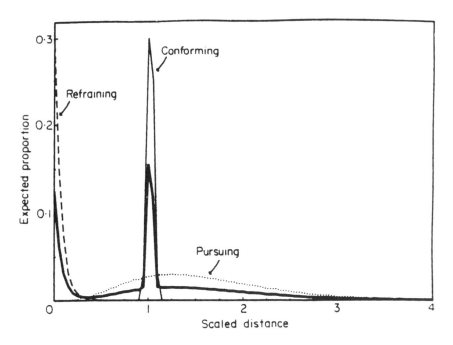

Fig. 5.14. A "component" distribution that matches an idealized mixed distribution (solid) line against the observed distribution of "refrainers," "conformers," and "pursuers" in lions (after Scheel and Packer, 1991).

suggest that this game may be best analyzed using the producer-scrounger model, first developed by Barnard and Sibly (1981).

5.7. Blood sharing and reciprocal altruism in bats

One of the earliest and most convincing studies of reciprocal altruism was Wilkinson's (1984) work on food (blood) sharing in vampire bats *(Desmodus rotundus)*. A typical group of vampires is comprised of females, who have an average relatedness of somewhere between 0.02 and 0.11 (Wilkinson, 1984, 1985b, 1990; but see Denault and McFarlane, 1995, for evidence that reciprocity occurs among male vampire bats). Females regurgitate blood meals to nestmates that have failed to obtain food in the recent past (Wilkinson, 1984, 1985a). Such food sharing can be a matter of life or death in vampire bats, as individuals will starve if they do not receive a blood meal every 60 hours (McNab, 1973). Wilkinson examined whether relatedness (measured via pedigrees), reciprocity (calculated using Wilkinson's "index of opportunity for reciprocity"), or some combination of the two best explained the evolution and maintenance of blood sharing in this species. The evidence indicates that:

> both relatedness and association predict regurgitation independent of their
> correlation. The relative effects of variables as measured by the standard-

ized regression coefficients . . . are approximately equal. These results indicate that regurgitation between distant relatives occurs only between animals that have been frequent roostmates as expected when reciprocity occurs. (Wilkinson, 1984, p. 182)

Beside the "index of opportunity for reciprocity," Wilkinson (1984) argues that three factors which suggest reciprocity are important in this system: (1) the probability of future interaction, w, is sufficiently high to select for reciprocity; (2) the blood meal obtained is critical, while the cost of giving up some blood may not be that great (satisfying one of Trivers's 1971 conditions for the evolution of reciprocal altruism); and (3) vampires are able to recognize one another and are more likely to give blood to those that have donated in the past (see Wilkinson, 1986, for evidence that one feature of social grooming in vampires is to facilitate individual recognition of cooperators and cheaters). Other cases of putative reciprocity in bats include work on information exchange (in the context of food calling) in the spearnose bat (*Phyllostomus hastatus;* McCracken and Bradbury, 1981) and the evening bat (*Nycticeius humeralis;* Wilkinson, 1992b) and cluster position in pallid bats (Trune and Slobodchikoff, 1978; but see Wilkinson, 1987, for alternatives to the first and last example, including hints that these examples may be closer to byproduct mutualism than to reciprocity).

5.8 Care-giving behavior in dolphins

Dolphins appear to meet Trivers' (1971) criteria for reciprocal altruism in that they are long-lived, intelligent (capable of individual recognition), and live in societies in which mutual dependence is great. Connor and Norris (1982) discuss three types of care-giving ("epimeletic") behavior (Scott, 1958) seen in dolphin species: (1) "standing by"—here individuals do not come to the assistance of an individual in need, but rather stay in their vicinity (Norris, 1958). This type of behavior often exposes the individual standing by to greater predation threat than it would experience if it left and returned to the group. (2) "assistance," which

> includes such behavioral sequences as approaching an injured comrade, showing violent or excited behavior in such circumstances, including interposition of the aiding animal between a captor and its prey, biting or attacking capture vessels and pushing an injured member away from a would-be captor. (Connor and Norris, 1982, p. 362)

and (3) "supporting" behavior, in which an animal supports a distressed individual at the surface, allowing it to breathe (Brown and Norris, 1956; Pilleri and Knuckey, 1969; Siebenaler and Caldwell, 1956; Caldwell et al., 1963).

In their review, Connor and Norris (1982) argue:

> Many aspects of the life of schooling dolphins have come to involve mutual assistance beyond those just discussed, and these often seem to be

related to reciprocity. In fact, the dolphin school and all its changing geometry becomes a system devoted to ordering reciprocity for its members. (p. 363)

Given this view, it is at first surprising that Connor and Norris provide not a single example (not even an anecdotal case) of reciprocal altruism among a pair of dolphins. The intent of their statement becomes a bit clearer however, when we realize that they are referring not to reciprocity between pairs, but rather to a "multi-party" system first envisioned by Trivers (1971):

> The multiparty model, in which individuals show generalized altruistic tendencies, we believe, does provide a means of explanation for dolphin altruism. In this case altruistic acts are dispensed freely and not necessarily to animals that can or will reciprocate. They need not necessarily be confined to the species of the altruistic individual. In human terms, a person can rescue a helpless fledgling that has fallen from a nest, and this does not imply any conscious intent toward the rest of society. The person is, instead, motivated first by a broad concept of distress and then by a complex of emotional responses, learning and social standards. To us, the evidence from dolphins clearly fits this model. The interspecific and intergeneric occurrence is explained. The somewhat unpredictable occurrence of altruism (some females abandoned, some males assisted) in dolphins also fits this model. (Connor and Norris, 1982, p. 370)

Whether such multi-party altruism is evolutionarily stable has not been the subject of any formal modeling (with the possible exception of Boyd and Richerson, 1989), nor has it been examined in controlled (or, for that matter, uncontrolled) settings.

5.9 Summary

Not surprisingly, cooperation is anything but rare among the nonprimate mammals. Individual recognition is common in this taxa, and plays no small role, at least with respect to the occurrence of reciprocity in this group. Reciprocity is, however, only one type of cooperation found in nonprimate mammals. In this chapter, I have outlined cooperation in the context of grooming behavior, alarm signals, coalitions, alloparenting, hunting, blood sharing, and epimeletic behavior. As always, examples are tied back to theory whenever possible.

6

Cooperation in mammals II: nonhuman primates

6.1 Introduction

Perhaps because we are so fascinated with cooperative behavior in our own species, this topic has been the study of extensive research in our closest living relatives—other primates. Attempting to write a chapter on cooperation in non-human primates is a daunting task. For example, in a recent edited volume on primate social behavior (Smuts et al., 1987), greater than 2000 references were cited. Of course, not all (or even the majority) were on cooperation, but the point is that the literature on this subject is huge, and I can at best make only a dent in it.

Because it would be easy to write a chapter on cooperation in chimps or vervets alone, I have made an effort to give examples from a wide variety of primate species. As in Chapter 5, it is difficult not to think of primates as somehow more "advanced" than nonprimates, especially because of the recent barrage of work on cognition in animals (e.g., Griffin, 1984; Harre and Reynolds, 1984; Else and Lee, 1986; Bryne and Whiten, 1988; Radner and Radner, 1989; Cheney and Seyfarth, 1990; Zentall, 1993). This gives us the chance once again to take a look, albeit a preliminary one, at the relationship between behavioral complexity and categories of cooperation.

6.2. Grooming behavior

Social grooming, or allogrooming, wherein one individual grooms another, is one of the most obvious and frequently noted cooperative behaviors recorded by primatologists. From the time of the earliest experimental work on primates, grooming has been considered a large part of the glue that holds primate troops together (Zuckerman, 1932; Carpenter, 1942). In general, dominant individuals receive more grooming than subordinates, who compete to groom their superiors (see Bernstein and Sharpe, 1966; Sade, 1972; Oki and Maeda, 1973; Rhine, 1973, for early evidence). In addition, individuals often groom those close to themselves in rank (Seyfarth, 1976; Cheney and Seyfarth, 1977; Chapais and Schulman, 1980; Seyfarth, 1980; Fairbanks, 1980). It is also worth noting that the theoretical (Seyfarth, 1977; Chapais and Schulman, 1980) and empirical work on the costs and benefits of grooming are heavily slanted toward the benefits.

As far as I can assess, few studies to date have experimentally examined the potential costs of grooming. Dunbar (1984a) found no significant "time-budget" costs associated with grooming in two species of baboons, *Papio anubis* and *Theropithecus geleda*. On the other hand, Maestripieri (1993) found that adult female macaques involved in social grooming were less vigilant with respect to the whereabouts of their offspring, thereby subjecting them to increased rates of aggression from other troop members (Figs. 6.1 and 6.2). In addition, anecdotal evidence from a number of species suggests that a potential indirect cost of grooming is paid by subordinates who defend their access to dominant individuals, and in so doing risk attack by others who wish to associate with the dominant.

Moving to the flip side of the coin, a partial list follows of the benefits associated with social grooming:

(1) Removal of ectoparasites. Perhaps the most obvious benefit accrued from grooming is hygienic, and many authors (Rosenblum et al., 1966; Sparks, 1967; Lawick-Goodall, 1968; Freeland, 1976; Hutchins and Barash, 1976; Kurland, 1977; Barton, 1985; Saunders, 1987) have suggested that removal of parasites is critical in understanding both the initiation and current utility of social grooming. Although there appears to be agreement on the importance of ectoparasite removal in the initiation of social grooming, Dunbar's (1991) compara-

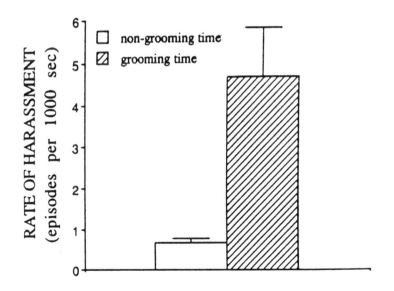

Fig. 6.1. Rate of harassment of individual macaques when mothers were involved in allogrooming versus when they were not (after Maestripieri, 1993).

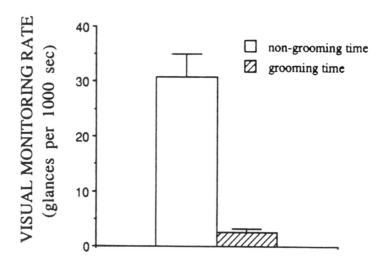

Fig. 6.2. Rate of maternal visual monitoring of infants when mothers were actively allogrooming versus when they were not (after Maestripieri, 1993).

tive study of 44 free-living species of primates suggests that grooming's current function is more social than hygienic. He hypothesized that if ectoparasite and hygiene-oriented aspects of grooming were important, then we might expect a positive correlation between body size and amount of time spent grooming. Conversely, if social aspects of grooming were paramount, a positive correlation between time spent grooming and group size should be apparent. The latter correlation was uncovered, but the former was not, suggesting that social aspects of grooming are of greater importance (at least in Old World monkeys; Dunbar, 1991).

(2) "Tension reduction." Terry (1970) has proposed that grooming serves the somewhat amorphous role of reducing "tension." This hypothesis has recently been tested (Boccia, 1987; Schino et al., 1988), and perhaps the most direct support of it comes from Keverne et al. (1989) who found an increase in beta-endorphin concentration in the cerebrospinal fluid of individuals that had been groomed.

(3) Coalition formation. One of the more popular hypothesized benefits to grooming is that the groomer/groomee relationship extends beyond this realm and into the formation of coalitions in many different contexts. I explore grooming and coalitions in somewhat more depth later in this chapter.

(4) Exchange for other currencies. In addition to a role in coalition formation (a triadic behavioral interaction), grooming has been hypothesized to play a role in a number of dyadic exchanges for other "currencies," such as "friendship" (Smuts, 1985); reduced aggression from other individuals, particularly dominant group members (Fairbanks, 1980; Silk et al., 1981; Silk, 1982); access to scarce resources such as water (Weisbard and Goy, 1976; Cheney, 1977) or food (de Waal, 1989; Kummer, 1978); access to a dominant's dominant

"friends" (Stammbach, 1978); entrance into new groups (Hauser et al., 1986); aid in chasing potential predators away (Kummer, 1978); and future association with individuals with "special skills" that others do not possess (Stammbach, 1988). It has also been suggested that males may have an increased probability of mating with the females they groom (Tutin and McGinnis, 1981; Smuts, 1985).

As is evident from above, grooming in primates can be used as an exchangeable currency. To avoid confusion, we need to adopt some terminology on this issue, first introduced by Hemelrijk and Ek (1991):

> Reciprocity and interchange are distinct. In reciprocity, for one kind of act, the same kind is received in return (e.g., grooming for being groomed). In interchange, however, two different kinds of acts are being bartered (e.g., grooming for the receipt of help in fights). (p. 923)

Note, however, that grooming can function in the realm of reciprocity and interchange at the same time, just as I can trade some dollar bills for different dollar bills, and others for yen. Before moving to specific examples of social grooming, it is important to be aware that there are a number of statistical problems inherent in measuring interchange and reciprocity (Seyfarth and Cheney, 1988). These problems have been addressed in depth by Hemelrijk and her colleagues (Hemelrijk, 1990a,b; Hemelrijk and Ek, 1991) and will be touched on in some of the examples provided below.

6.2.1 Reciprocal grooming in chimpanzees

Using the now famous chimp colony at Burger's Zoo in Arnhem, The Netherlands (see de Waal, 1982, for more on this group), Hemelrijk and Ek (1991) examined whether grooming was reciprocated in both female and male chimpanzees *(Pan troglodytes),* both in the absence and presence of a clear alpha individual in a group (Fig. 6.3). In order to avoid spurious correlations with rank and other variables in their analysis, Hemelrijk and Ek used partial correlation statistics developed for studying reciprocity and interchange (Hemelrijk, 1990a,b). Given the cognitive complexity of chimps (e.g., the possibility that they may practice "ostracism" in the wild; Goodall, 1986a), it is not surprising to learn that males and females reciprocated grooming acts in both the absence and presence of a dominant group member. The situation is, however, much more complex with respect to the relationship between grooming, coalition formation, and sex (Hemelrijk and Ek, 1991; see below for other examples of such complexity).

6.2.2 Grooming and coalitions in vervets

As is evidenced in their splendid book, *How Monkeys See the World* (Cheney and Seyfarth, 1990), field work on vervet monkeys *(Cercopithecus aethiops)* has proven to be a treasure chest of information on cooperation in primates. For example, unrelated vervets seem to exchange grooming for aid in coalitions, while no such exchange exists among related individuals (Seyfarth and Cheney, 1984; Figs. 6.4 and 6.5). Seyfarth and Cheney (1984) uncovered these

Fig. 6.3. Reciprocal grooming among a pair of chimps at The Arnhem Zoo in the Netherlands. (Photo courtesy of C. Hemelrijk.)

fascinating results while conducting field tests in Amboseli National Park in Kenya. Armed with the information that vervets are able to recognize each other's calls (Cheney and Seyfarth, 1980, 1982), they chose two individuals, A and B, and taped the vocalization used by A to solicit aid from B. This tape was then played near individual B, either after a grooming bout between A and B, or when no such bout had taken place. The degree of relatedness between A and B was also a treatment in this experiment, with individuals classified as either kin ($r \geq 1/4$) or nonkin ($r \leq 1/8$).

When no grooming took place between A and B, kin responded more strongly to solicitation calls than nonkin, as might be expected from inclusive fitness theory. However, prior grooming had no effect on response to solicitation calls among kin, but had a significant effect among nonkin (Fig. 6.6). Nonkin therefore seem to exchange grooming for aid, while kin do not. Seyfarth and Cheney (1984) suggest that the lack of a grooming treatment effect among kin may be due to the fact that kin always respond strongly to one another, and this may mask any treatment effect from grooming. Hemelrijk (1990a) and Hemelrijk and Ek (1991), question, however, whether even nonkin in vervets are in fact engaging in any interchange. Using statistics designed for statistically analyzing reciprocity and interchange, Hemelrijk and Ek (1991)

Fig. 6.4. Allogrooming in the vervet monkeys of Amboseli National Park, Kenya. (Photo courtesy of D. Cheney.)

Fig. 6.5. Coalition formation in vervet monkeys (Amboseli National Park, Kenya). (Photo courtesy of D. Cheney.)

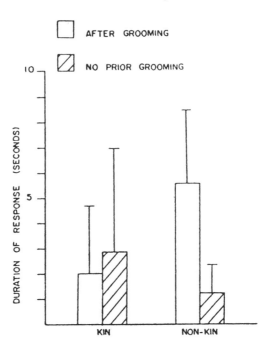

Fig. 6.6. Response of vervets to kin and nonkin recruitment vocalizations in the presence and absence of prior grooming bouts (after Seyfarth and Cheney, 1984).

argue that Seyfarth and Cheney's results are not evidence of interchange per se. Rather than a true correlation existing between *being* groomed and coalition support, each of these variables is independently correlated with the act of grooming, and when this is statistically removed from their data, the correlation between grooming and coalition aid disappears (Hemelrijk, 1990a; Hemelrijk and Ek, 1991). Nonetheless, as Hemelrijk (1991) acknowledges, Seyfarth and Cheney's (1984) ingenious field work on cooperation in vervets is just the direction in which primate ethologists should be heading.

6.2.3 Grooming, reciprocity, and food exchange in chimpanzees

Food sharing among chimpanzees is well documented in both the laboratory and the field (Fig. 6.7; see Tuttle, 1986, for a review). A number of hypotheses have been put forth to explain this type of cooperative behavior. de Waal (1989) categorizes these hypotheses as follows: (1) the "sharing-under-pressure" hypothesis: first advanced by Wrangham (1975) and Blurton-Jones (1987), this hypothesis argues that chimps use food to "pay" others not to be aggressive toward them; (2) "the share-to-enhance-status" hypothesis: as implied by the name, this hypothesis supposes that chimps share food to increase their standing within groups (Rijkesen, 1978; de Waal, 1982; Moore, 1984); and (3) the reciprocity hypothesis: here, de Waal (1989) argues that chimps give food to those who have provided them with either food or some other currency in the past. De Waal (1989) constructed an experiment to examine all three hypotheses in the chimp. Based on thousands of observations, he found evidence supporting hypotheses 1 and 3. In conjunction with "the sharing-under-pressure" hypothesis, he found that individuals with low rates of food distribution received more aggression than did individuals with higher rates. Chimps, however, were also reciprocating acts, in that individuals receiving food were likely to return the favor. However, no evidence for anything like TFT behavior was found. Evidence for the "interchange" of grooming and food was also uncovered. If A groomed B in the morning, then B was more likely to give food to A that afternoon. Furthermore, if A groomed B, A was unlikely to give B food in the afternoon (de Waal, 1989).

6.2.4 Grooming and associations with "specially skilled" individuals in macaques

While grooming can in principle be exchanged for any good, it is particularly interesting when it is exchanged for nonmaterial goods. In a fascinating experiment, Stammbach (1988) trained all members of a long-tailed macaque *(Macaca fascicularis)* group how to use a simple lever machine to obtain food (popcorn). The operation of the machine was then made more complex (a three-lever operation), and only a single, subordinate group member was trained to complete this task. Once the task was completed, however, other group mem-

Fig. 6.7. Food sharing between three chimps at the Arnhem Zoo in the Netherlands (the female on the right is in estrous condition). (Photo courtesy of C. Hemelrijk).

bers were able to feed at the dispenser. Although subordinate specialists did not rise in rank, they received significantly more grooming than before they were trained, and this increase in affiliative behavior took place away from the feeding machine. It seems, therefore, that long-tailed macaques are able to recognize specially skilled individuals in their midst, and exchange grooming for the opportunity to spend time with such an individual, perhaps to learn the skill or perhaps simply to obtain food when the trained individual provides the opportunity to feed.

6.2.5 Grooming and kinship

Kinship is at the core of much of primate behavior (e.g., Morin et al., 1994). Primates are able to recognize kin (Gouzoules, 1984; Gouzoules and Gouzoules, 1987), and within the context of grooming, kinship often plays an important role (Gouzoules, 1984; Gouzoules and Gouzoules, 1987). As noted above, individual primates often groom those close to them in the hierarchy. It turns out, however, that closely ranked individuals, because of the "inheritance" of maternal rank (Cheney, 1977; Datta, 1983; Lee, 1983; Horrocks and Hunte, 1983; also see Chapais, 1992, for a review), are also related, and so kinship may play a critical role in grooming behavior (Seyfarth, 1976; Seyfarth, 1980; Fairbanks, 1980). In some cases, kinship and reciprocity are both involved in cooperative grooming. For example, Nishida (1988) found that in wild chimpanzees, mothers groomed offspring from birth, but offspring rarely groomed parents until they reached the age of about four (female offspring began grooming a bit earlier than males). By age eleven, however, Nishida's (1988) reciprocity index indicated that offspring and parents were engaged in true reciprocal grooming.

Silk et al. (1981) examined the effects of sex, rank, and kinship in adult and juvenile bonnet macaques *(Macaca radiata)*. Not surprisingly, they found that "Kinship, rank and sex have complex and variable influences upon the relationships of females and immature bonnet macaques" (p. 129). Nonetheless, kinship emerges as perhaps the most salient feature of grooming relationships among adults and young. For example, even though adult females groom the daughters of high-ranking females more than they are groomed by such individuals (indicating an important role for rank in grooming relationships), females, overall, groom related immature individuals more often than unrelated immatures. In addition, females are much less likely to direct acts of aggression toward related versus unrelated young (see Silk, 1982, 1992a, for more on kinship and coalitions in bonnet macaques and Silk, 1992b, for more on grooming in exchange for support). This kin-based grooming may have major ramifications on lifetime reproductive success as it is tied to the inheritance of maternal rank (Silk et al., 1981), which is itself correlated with reproductive success (Kauffman, 1965; Drickamer, 1974; Hausfater, 1975; Sade et al., 1976; Dittus, 1979; Packer, 1979; Dunbar, 1980; Silk et al., 1980; Dixson et al., 1993; but see Dunbar and Dunbar, 1977; Gouzoules et al., 1982; Altmann et al., 1988; Bercovitch, 1991, and Packer et al., 1994, for contrary opinions).

6.3   Nongrooming-based coalitions

Primate social biologists have become enamored with studying coalitions in a wide variety of contexts, and in so doing have built a number of mathematical models of coalition formation constructed to understand both the proximate and ultimate reasons underlying this complex behavior (e.g., Aoki, 1984; Boyd and Richerson, 1989; Noë, 1990, 1992; Noë et al., 1991; Boyd, 1992). Below I highlight some fascinating cases of cooperation and coalition formation in primates. For a more general overview of this subject, Harcourt and de Waal's (1992) edited volume *Coalitions and Alliances in Humans and Other Animals* is indispensable.

6.3.1   Coalitions and reciprocity in Papio anubis

Perhaps the most well-known study of coalitions among primates, and indeed one of the most cited papers on reciprocity in general, is Packer's (1977) now classic work on male reproductive coalitions in baboons, *Papio anubis.* One of the aims of this study was to separate kinship from reciprocity, and while Packer could not be sure of the exact pedigrees of all individuals, known bloodlines and demographic data (males always emigrate from their home troop; Packer, 1975) suggest that coalition partners were *unrelated.*

P. anubis males actively solicit coalition partners, with soliciting being

> a triadic interaction in which one individual, the enlisting animal, repeatedly and rapidly turns his head from a second individual, the solicited animal, towards a third individual (opponent) while continuously threatening the third. The function of headturning by the enlisting animal is to incite the solicited animal into joining him in threatening the opponent. (p. 441)

Packer (1977) observed 140 solicitations, 97 of which resulted in coalitions. On 20 of these 97 occasions, the opponent was consorting with an estrous female, and this had the effect of increasing the animal's probability of entering into a coalition. On six occasions, the estrous female left the opponent and went to the enlisting male. This created a clear-cut benefit of coalition formation for enlisting individuals. From a solicited animal's perspective, however, joining a coalition may be costly in that such individuals never obtained the estrous female, but risked some probability of attack from the opponent while the solicitor was with the female. The obvious question then is, Why join when solicited? Packer's study suggests that the answer may be related to the fact that (1) those who joined coalitions also received aid most often when they solicited, and (2) opponents rarely fought against pairs of males in a coalition. As such, the enlisted individual's cost was probably low, but the payoff (an estrous female) when soliciting in the future was quite large. However, as Packer notes, just because those who joined coalitions were most likely to have their solicitations responded to is not prima facie evidence for reciprocity. That is, in order to document reciprocity, one must show more than just a broad correlation between the number of times one joins a coalition and the number of times one is joined by others in coalitions. Rather, direct evidence is needed to show that specific pairs or individuals are involved in relationships within which such a correlation exists. Packer found just such a relationship in that individuals had "favorite" partners, and that favorite partners solicited each other more often than they solicited other group members (Table 6.1).

Subsequent work on coalitions in baboons has generally confirmed Packer's findings (Smuts, 1985; Noë, 1986; but see Noë 1990). Bercovitch (1988), however, found that in his study of olive baboons *(Papio cyanocephalus anubis),* males who solicited coalitions were no more likely to obtain the estrous females than any other coalition member, and males who refused to join a coalition were again solicited in the future. The reason for the discrepancy between Packer (1977) and Bercovitch's (1988) studies are not clear, but Hemelrijk and Ek (1991) suggest that it may be tied to the presence of a clear alpha male in the latter, but not the former study.

6.3.2. Reciprocal alliances and behavioral plasticity in baboons

In her study of coalition formation among free-ranging baboons *(P. cyanocephalus ursinus),* Cheney (1977) uncovered a curious case in which alliance formation shifts from a nonkinship- to a kinship-based system. When adult baboons form coalitions among themselves, Cheney found that it was typically with *relatives* that held a similar rank. Kinship played only a minor role, however, in coalition formation among immatures. Instead, immatures attempted to form coalitions with the "best" (i.e., most powerful) partners, namely the offspring of high-ranking individuals (see de Waal, 1990, for evidence that mothers may even encourage such relationships in some primates). However, as immatures age and high-ranking coalition partners often fail to reciprocate when needed, baboons shifted toward kinship-based coalitions.

Table 6.1. Soliciting behavior of favorite partners of target males (after Packer, 1977).

Target	No. of occasions favorite partner solicited target	Mean no. of occasions favorite partner solicited each male	Favorite partner solicited target more (+) or less (−) than average
BBB	3	2.7	+
CRS	3	1.2	−
DVD	0	1.0	+
EBN	5	1.0	+
GRN	1	0.7	+
JNH	2	1.1	+
LEO	4	2.7	+
MNT	1	0.4	+
WDY	5	2.7	+
WTH	4	2.8	+

6.3.3.   Symmetry-based reciprocity, calculated reciprocity, and revenge systems

In his thought-provoking book *Chimpanzee Politics,* de Waal (1982) provides an animated account of the complex web of social interactions that defines chimpanzee group life, and speculates on what this may tell us about primate cognition. De Waal and Luttrell (1988) follow up this line of inquiry by examining "symmetry-based reciprocity" and "calculated reciprocity" in chimps, rhesus monkeys *(Macaca mulatta),* and stump-tailed macaques *(Macaca arctiodes).* One example of symmetry-based reciprocity might be interindividual association patterns (de Waal and Luttrell, 1988; also see Dugatkin and Sih, 1995, for more on this and similar phenomena), wherein *A* spends most of its time near *B,* and by default *B* spends a good deal of time near *A.* In this case, exchange of behaviors might simply be due to proximity rather than some type of score keeping. De Waal and Luttrell (1988) outline the distinction between their two types of reciprocity:

> We need to distinguish *symmetry-based reciprocity* from *calculated reciprocity.* The first type involves exchanges between closely bonded individuals who help each other without speculating equivalent returns. . . . This reciprocity mechanism should be assumed in relationships with symmetrical characteristics—such as those between kin and frequent associates—unless it can be excluded. . . . The second type of reciprocity is calculated by feedback, that is, the continuation of helpful behavior is contingent upon the partner's reciprocation. An intimate relationship is not required. Whether non-human primates possess the cognitive capacities necessary for this more complex scheme is not yet certain, although both anecdotal (de Waal, 1982) and experimental evidence (Savage-Rumbaugh et al., 1978; Lefebrve, 1982) suggest a concept of trade in chimpanzees. . . . (p. 103, emphasis added in original)

de Waal and Luttrell searched for both types of reciprocity in the context of agonistic interactions in chimps, rhesus monkeys, and stump-tailed macaques (also see de Waal, 1992). After removing "symmetry features," they found evidence for calculated reciprocity in all three species—individuals intervened *on* behalf of an individual who intervened on their behalf (de Waal and Luttrell's "pro-intervention"; Fig. 6.8). However, only chimpanzees followed a reciprocity rule in the context of interventions *against* those who intervened against them, and this was found only when the chimps were tested in a large open area in the Arnhem Zoo. de Waal and Luttrell interpret this "contra-intervention" as evidence of a cognitively sophisticated *revenge* system in chimps (but see Hemelrijk and Ek, 1991) and suggest that pro-intervention may be the phylogentically primitive case, and contra-intervention the more advanced trait.

Following up experimental evidence that male bonnet macaques *(M. radiata)* reciprocate aid in coalition formation (see Silk, 1992a, for a review), Silk (1992b) uncovered the first case of "revenge" and "loyalty" in monkeys. Based on 550 observations, she found that: (1) Kinship played a role in coalition

Fig. 6.8. Coalition formation among chimps at the Arnhem Zoo in the Netherlands. (Photo courtesy of F. de Waal.)

formation in this species. Even though most males in the study group were unrelated, males were more likely to intervene on behalf of maternal kin than unrelated individuals. Such intervention was cooperative in that "(a) the beneficiaries of support reduced the probability of being defeated when they and their supporters outnumbered their opponents and (b) supporters were sometimes intimidated when they intervened in ongoing disputes." (p. 320). (2) Although males tended to intervene against most others in their group at one time or another, and tended to avoid intervening against high-ranked individuals, Silk did find evidence for a revenge system in that monkeys were most likely to intervene against those who intervened against them. (3) Males displayed "loyalty" to some partners, that is, they were likely to support an individual *and* not intervene against them, and even though there was considerable variation among males in their tendency to be loyal, males were consistently more loyal toward kin and toward those who had been loyal to them.

Silk's results with respect to revenge and loyalty cast some doubt on the claims that monkeys may be less cognitively complex than chimps, and that pro-interventions are a primitive trait when compared with contra-interventions (de Waal and Luttrell, 1988). Hopefully, primate social biologists will follow de Waal and Luttrell's and Silk's lead and address these issues in other species of monkeys, as well as apes.

6.3.4 Coalitions, kinship, and the inheritance of rank in gorillas

As mentioned earlier, "inheritance" of maternal rank is not uncommon in primates. The mechanism producing such an inheritance system, however, is controversial (Datta, 1992). Harcourt and Stewart (1987) propose a three-stage process by which rank might pass from parent to offspring: (1) offspring that are vulnerable are protected when attacked; (2) immatures who successfully receive such aid initiate aggressive contests against others who are not as protected by their kin; and (3) dominance rank so attained is maintained by continued support. Harcourt and Stewart (1987) tested this hypothesis on the gorilla, *Gorilla* (or *Pan*) *gorilla,* because the inheritance of maternal rank is *absent* in this magnificent species (Fig. 6.9). As such, one can examine which part of the three-step process is missing in the gorilla, and thus indirectly test this hypothesis.

When studying two wild gorilla populations in Rwanda and Zaire, Harcourt and Stewart (1987) found that adults intervened on behalf of juveniles about 14% of the time (similar to that found in other primate studies). In support of stage 1, almost all interventions were on behalf of immature kin. Furthermore, young kin were supported over older kin and close kin over distant kin. No evidence for stages 2 or 3, however, was found in gorillas. Although support often amounted to juveniles obtaining a contested resource, dominant mothers were not more likely to help kin than subordinate mothers. Why do females fail to support their young, when such support may be helpful? Harcourt and Stewart (1989) suggest three possibilities. First, it may be that the resources obtained by juveniles (low quality,

Fig. 6.9. An adult female gorilla (far left) cuffs a juvenile male (to her right). The juvenile male's mother (above the tree), grandmother (right of tree), and aunt (small individual at fore right) aggressively intervene on behalf of the juvenile male. As a result, they supplant the unrelated female who initiated aggression toward their kin and gain sole access to the tree (after Harcourt, 1992). (Photo courtesy of A. Harcourt.)

widespread plants) when supported were not worth the costs of such support. Second, the great difference in the age of immatures in the study group made support too costly. Third, the dominant male, who fathered most of the group members, may simply have not permitted the unequal distribution of resources to his offspring (a likely outcome of coalition formation).

6.4 Cooperative hunting

If all the accounts of coalitions and aggression have yet to shatter any myths you might have had of the "friendly chimp," depictions of cooperative hunting in this species, in all likelihood, will. Cooperative hunting or the lack of it has been examined in chimp populations in the Gombe Preserve (Tanzania; Lawick-Goodall, 1968; Telecki, 1973; Busse, 1978; Goodall, 1986b; Boesch, 1994a, b, c), the Mahale Mountains (Tanzania; Nishida et al., 1983, 1992; Uehara et al., 1992), the Tai National Park (Ivory Coast; Boesch and Boesch, 1989; Boesch, 1994b) and, in a stretch of the word "hunting," even in the lab (in the sense of cooperating on a learning task to obtain food; Chalmeau, 1994). Following Telecki's (1973) discussion of "active cooperation" in chimp hunting behavior at Gombe, Busse (1978) examined cooperative hunting in our closest living relative. Busse studied chimp hunts involving red colobus monkeys *(Colobus badius),* a small (1 to 4 kg) arboreal folivore (Boesch 1994a). Although meat may make up as little as 1 to 3% of a chimp's diet (Wrangham, 1975), red colobuses make up the majority of such prey for the Gombe population (Wrangham, 1975; Busse, 1978; Goodall, 1986b). Based on data collected from 64 hunts, Busse found no evidence for cooperative hunting (Fig. 6.10). In par-

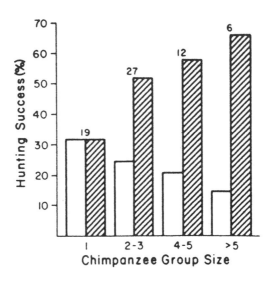

Fig. 6.10. Hunting success in chimps as a function of hunting group size. Hatched bars show the success of chimp groups and open bars show the success of average individuals in a group (after Busse, 1978).

ticular, chimp capture rates were no higher in pairs than when hunting alone. As predicted by Packer and Rutton's (1988) cooperative hunting model (reviewed in Chapter 5), when it comes to hunting small prey like red colobus monkeys, it seems that chimps at Gombe just hunt when they happen to be in groups, rather than truly hunt cooperatively.

The most comprehensive studies of cooperative hunting among chimps are those of Boesch and Boesch (1989) and Boesch (1994b,c), who compared hunting patterns across the Gombe and Tai Chimp populations (Fig. 6.11). Following work that indicated that cooperative hunting was more common among Tai chimps (Boesch and Boesch, 1989), Boesch (1994b) constructed a game theory model and tested its predictions on chimps from both populations. In this model, two strategies, hunter and cheater, compete with one another. Cooperative hunting emerges as a solution to the game when the relationship between group size and hunting success are nonadditive. Further, cooperative hunting is stable (cannot be invaded by cheaters) in this model only when a social mechanism exists to prevent cheaters from sharing equally on captured prey.

Major differences in hunting strategies emerged between the Gombe and Tai populations (Boesch, 1994b). In Tai chimps, hunting success was correlated with group size in a nonadditive fashion, indicating ". . . a synergism between the individual hunters, which probably results from the better organization of the individuals when hunting in larger groups" (Boesch, 1994b, p. 657).

Furthermore, a very complex, yet subtle social mechanism exists which regulates access to fresh kills and assures hunters greater success than cheaters. As such, byproduct mutualism has produced a stable system of cooperative hunting in Tai chimps.

The situation was quite different at Gombe. There, the success rate for solo hunters was quite high (compared with the Tai population), and no correlation between group size and hunting success was found. In addition, behavioral mechanisms limiting a cheater's access to prey were absent, and cheaters received as much food as cooperators (Goodall, 1986b; Boesch, 1994b). Despite this, however, Boesch found that 52% of all hunts were conducted in groups. Why?

> A precise look at the behaviour of hunters may provide an explanation. Gombe chimpanzees, when hunting in groups, start to hunt on the same group of prey, but as a rule, each follows a different target prey (Busse, 1978; Goodall, 1986a; Boesch, 1994a) and they do not coordinate their movements. Thus, group hunting at Gombe is better described as simultaneous solitary hunts than true cooperation. . . . (Boesch, 1994b, p. 663)

As in Busse (1978) and Packer and Rutton (1988), this quote highlights the danger of necessarily equating hunting that occurs when individuals happen to be in groups with cooperative hunting.

6.5 Intergroup conflicts and cooperation

Although much of the primate social literature is focused on the behavior of individuals within groups, some of the most interesting action occurs during

Fig. 6.11. Meat sharing in the cooperative hunting chimps of the Tai forest. (Photo courtesy of C. Boesch.)

between-group encounters. Such encounters on occasion can be friendly, and even solicited (e.g., Cheney, 1987; Judge, 1994), but most often they are not. Here I will concentrate on how between-group *conflict* affects within-group *cooperation*.

6.5.1 Raiding and "warfare" in chimpanzees

As Boehm (1992) notes, "warfare" is often defined as large-scale, open hostility between groups, in which both sides in the conflict use lethal force against the other. According to this definition, chimpanzees do not engage in war. Yet between-group interactions often appear to be "war-like," and in fact resemble the "raiding" behavior so common among many tribes of humans (Boehm, 1992; Figs. 6.12 and 6.13). In humans, "a raid is conducted into enemy territory by a relatively weak force which relies on coordination, stealth, and an element of surprise, either to quickly kill a few of the enemy, or to quickly take away readily transportable commodities. . . ." (Boehm, 1992, p. 153). As is often the case, cooperation in this example is tied to competition. That is, cooperation is critical within groups, in order for a group to emerge victorious over its rival. Although the costs and benefits of raiding behavior are not known (e.g., is it truly dangerous to be in a raiding party?), and the behavior itself has not been examined in controlled experiments, it appears to be a good example of cooperation via group selection. Selection between groups favors such raids (as they likely benefit all group members, not just the raiding party participants), yet, if raiding is dangerous, selection within groups should favor cheating—that is, letting others do the raiding, but continuing to reap the benefits.

During raids, all-male chimpanzee "patrol" groups often travel into areas that border their territorial boundaries (Bygott, 1979; Goodall, 1986b). In contrast to "excursions" for food, in which vocalizations are common, patrols move in a wary fashion and remain silent (Goodall, 1986b), in accordance with the above definition. These raids often involve the killing of a small number of members of the raided group, and the capture of females. Occasionally raiding parties from two groups will meet one another. Rather than all-out aggression, both groups engage in hostile vocalizations and then withdraw (Goodall, 1986b). Despite the fact that all-out warfare does not emerge when two raiding parties meet, raiding can, in the long run, amount to the extinction of one group. Nishida et al. (1985) provide evidence that raiding behavior in the Ma-

Fig. 6.12. "Warfare" prelude—chimps listen to the calls of strangers from across the valley (Gombe Forest, Tanzania). Photo provided by Anthro-Photo.

Fig. 6.13. Chimp band "calling" toward opposition band during "warfare" (Gombe Forest, Tanzania). Photo provided by Anthro-Photo.

hale Mountains of Tanzania amounted to a larger group extinguishing a smaller group of chimps (also see Goodall, 1986b, for evidence of this at Gombe).

6.5.2 Intergroup cooperation and intergroup competition: comparative studies of within- and between-group selection

Van Schaik (1989) has argued that when between-group competition is strong, within-group cooperation in female-bonded primate societies should be common, producing more "egalitarian" societies (Vehrencamp, 1983; Hand, 1986). This should be particularly true in species which show a positive correlation between group size and individual reproductive success (see Wrangham, 1980, 1987, for more on this). In part to test the above predictions, Isbell (1991) undertook a comparative review of 20 female-bonded primate species. She found no evidence that the strength of dominance hierarchies correlated negatively with the intensity of between-group selection. Cheney (1994) points out, however, that it is possible to have dominance hierarchies, and still have some egalitarian relationships, and so the exact relationship between within- and between-group selection and inter-group competition remains unknown.

In order to better understand this relationship, Cheney (1994) examined the distribution of grooming behavior within groups and the degree to which females were involved in between-group competition. This was done using field and laboratory data on vervets and published data from 29 other primate species. When comparing species characterized by female philopatry versus female dispersal, Cheney found support for Isbell's (1991) hypothesis in some contexts. However, in perhaps the most important analysis—female-bonded species that live in socially differentiated groups—the hypothesis was not supported, as females were no more diverse in their grooming habits when between-group selection was strong. Cheney suggests two reasons for this last result. First, when between-group selection is strong, it may be more important for females to strengthen the bonds they already have, rather than form many new bonds. Second, dominant females may attempt to exploit subordinates and provide only enough group life benefits to make it worth remaining in a group. This might manifest itself in the lack of an egalitarian distribution of grooming. More data is needed to differentiate these possibilities, as well as to truly understand the relationship that exists between group selection and intergroup aggression in primates.

6.6 Scanning, alarm calls, and other anti-predator behaviors

6.6.1 Scanning "roles" in red-bellied tamarins

In addition to emitting particular vocal signals in the presence of predators (Moynihan, 1970), red-bellied tamarins *(Saguinus labiatus)* spend a significant proportion of their time "scanning" the environment (Caine, 1984). Using a laboratory colony of *S. labiatus,* Caine (1984) examined the function of such scanning. In a series of experiments, she found that scanning was most common during the most dangerous times of the day, and that tamarins scanned much more often when a "dangerous" stimulus was presented than when a "neutral" stimulus was presented. But the most interesting aspect of this study is that individuals may partition scanning duty among group members (also see Hall, 1960; Altmann, 1979; Rhine, 1981, for further discussion of scanning "roles" in primates). Caine (1984) notes that

> when one looks across time periods within subjects, it is clear that, in most cases, each of the tamarins had one particularly active scanning time. For example M1 was the most active scanner in group 1 at lights on; he was only the 6th most active during the early morning session. Conversely, M2 was the second most active group 1 scanner in the later afternoon, but only the 6th most active at lights on. The origins of these differentiations are unknown. . . . (p. 61)

If scanning includes costs to the scanner, then red-bellied tamarins appear to have devised a system of dividing up such costs. Whether or not this is a case of reciprocity remains unanswered, as how tamarins respond to an individual failing to scan at the appropriate time remains unknown (see Caine et al., 1995, for more on the similarities between tamarin *food* calls and sparrow "chirrup" calls; Elgar 1986, see Chapter 4).

6.6.2 Kinship and alarm calls in Kloss's gibbons

Kloss's gibbons *(Hylobates klossii)* are found off the west coast of Sumatra, Indonesia. Tenaza and Tilson (1977) looked at two types of long-distance alarm calls made by these gibbons: "sirening" and "alarm trills." Hunting by humans is by far the predominant predation force operating on *H. klossii,* and data indicate that hunters stop their pursuit of a gibbon once one of these alarm calls is made.

Tenaza and Tilson (1977) postulate that kinship also plays a role in this system. They found that most of the time, an individual's mate and offspring were found within 10 m of the alarm caller, yet alarm calls were audible for at least 800 m. Is it really necessary to give such long-distance alarm calls when immediate family are usually found so close? The answer may lie in the demographics of territory settlement. With an admittedly small sample, Tenaza and Tilson (1977) found that offspring often settle in territories adjacent to their parents. Given that territories are about 400 m long, long-distance alarm calls would then reach offspring in neighboring territories, providing a kinship-related benefit to such calls.

6.6.3 Intentionality and alarm calls

Alarm calling in vervets is serious business. In fact, because of all the information that such calls provide about monkey perceptions and attributions, a fair share of Cheney and Seyfarth's (1990) book *How Monkeys See the World* is devoted to alarm calls in vervets and other monkeys. In vervets, for instance, very distinct and different calls are made in response to leopard, snake, and eagle predators (Struhsaker, 1967a,b). Whether the monkeys actually use specific alarm calls to refer to specific predators was the subject of some debate (Marshall, 1970; Smith, 1977). It is possible, for instance, that different alarm calls reflect differences in emotional states, or that alarm calls just attract attention, and that individuals respond to a predator in a specific way only after it has been seen. While it is true that the length of alarm calls can be ranked: leopard > eagle > snake, thus providing some support for the "arousal" hypothesis, Seyfarth et al. (1980a,b) provide evidence that vervets respond in adaptive ways to different calls—for example running to the trees when a leopard call is heard, or hiding in bushes when an eagle alarm call is uttered. This suggests that vervets are at least using calls to refer to the *mode* of predator attack (see Macedonia, 1990, and Evans et al., 1993, for more on this in ring-tailed lemurs, *Lemur catta,* and chickens, *G. gallus,* and Macedonia and Evans, 1993, for a recent review). Furthermore, vervets can learn the alarm calls of other species (e.g., birds; Hauser, 1988), and even use "fake" calls in order to cause other troops of vervets, that may pose a threat, to disperse (Cheney and Seyfarth, 1990).

The list of interesting aspects of vervet calls is almost endless. With regard to cooperation, however, a very tricky question can at least be addressed in this species. That is, do alarm callers *intend* to warn others of potential danger?[*]

---

[*]I do not mean to imply that intentionality (sensu Dennett, 1987) need be part of our definition of cooperation. If an action affects others, in some respect, intentionality is irrelevant to its evolution—only the costs and benefits of the action matter (see Wilson, 1990b; Wilson and Dugatkin, 1992, for discussions on psychological and evolutionary definitions of altruism). Nonetheless, given that alarm calling is a dangerous activity, the intentionality of alarm callers may help us better understand the evolution of cooperative behavior.

Cheney and Seyfarth (1990) address the intentionality of alarm calling with data from vervets and Japanese macaques *(Macaca fuscata)*. In support of the notion that alarm callers *intend* to notify those who might hear the call, they found that lone vervets do not give alarm calls. Furthermore, vervets give more alarm calls when their offspring are in the vicinity than when the offspring of others are present, and males give more calls when a female, as opposed to a dominate individual, is nearby (Cheney and Seyfarth, 1985). Both the failure of lone individuals to call and context specificity are necessary for intentionality, but are not unique to primates (see Cheney and Seyfarth, 1990, for other examples in birds and nonprimate mammals). A stronger test of intentionality would examine whether the state of those potentially being warned affected the alarm callers' behavior. For example, if alarm calls are intended to warn others in a group, such calls need not be emitted if the caller recognizes that its groupmates are already aware of any putative dangers. In a similar vein, once others in the group hear an alarm call, the individual emitting it should stop. Field data indicate that vervets continue to give alarm calls after everyone in the group has heard them (Cheney and Seyfarth, 1981, 1985). Before, however, ruling out intentional alarm calls, Cheney and Seyfarth (1990) ran a controlled experiment on alarm calls and intentionality in a laboratory colony of Japanese macaques. In this experiment, mother-daughter pairs were tested in two treatments "knowledgeable" and "ignorant." In the "knowledgeable" treatment, both mother and daughter sat next to each other, and could see a "predator" (in this case a human with a net). In the "ignorant" treatment, only the mother could see the predator. If mothers intended to inform their daughters of danger, then alarm calls and other protective behaviors should have been greater in the "ignorant" treatment. This, however, was not the case, and the evidence did not support the notion that mothers intended to warn offspring (Fig. 6.14). Despite

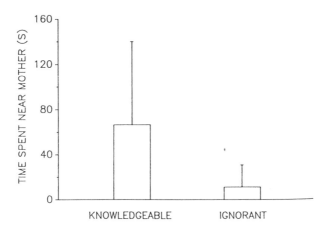

Fig. 6.14.   The number of seconds spent by "knowledgeable" and "ignorant" Japanese macaque juveniles within an arm's reach of their mothers (after Cheney and Seyfarth, 1990).

the failure to uncover intentionality in macaques, this type of experimental protocol will no doubt be very useful in future work on cooperation in primates.

6.7 Alloparenting and helping

In group-living primates, individuals have an acute interest in newborn and young troup members. While mothers are often reluctant to let their babies interact with other group members at too early an age (Nicolson, 1987), alloparenting is found in many primate species (see Hrdy, 1976; Silk, 1980, for a discussion of the potential costs of alloparental care). Although many similarities exist between alloparenting in primates and nonprimates, Fairbanks (1990) notes:

> Primate allomothering differs from cooperative breeding in birds and carnivores in several respects, however. Although there are exceptions (Pereira and Izard, 1989), primate allomothers typically do not provide food for the young. Many primate allomothers are immature animals who are not yet physiologically capable of breeding. . . . (p. 553)

6.7.1 Aunts and infants in rhesus monkeys

Rowell et al. (1964) provide the first detailed description of the "aunt"-infant relationship in laboratory rhesus monkeys *(Macaca macaca).* In rhesuses, "aunting" behavior starts early on, as Rowell et al. (1964) observed "aunts"— who themselves were childless—close by the side of a pregnant female in two live births (and in three of the six births that were not observed directly, the individual who gave birth was "attended" to by a female within three hours). When infants were very young, aunts would approach a mother cautiously when attempting to spend time near the child. Aunts were often aggressive to others who attempted to approach them when they were with a baby, and once infants grew older, aunts occasionally would "cuff" or gently threaten a very active baby they were watching. Albeit rare, "baby-sitting" behavior did occur, with aunts watching an infant while its mother was foraging.

Although seminal to the study of aunting behavior in primates, Rowell et al.'s (1964) work was observational, rather than experimental, making interpretations of aunting behavior tentative at best. For example, kinship was not discussed in this paper ("aunt" was used in the generic sense), and so it is impossible to know whether kin selection played any role in who displayed aunting behavior, and to what extent they displayed it. However, Rowell et al. did provide one clue with respect to cooperation and aunting behavior. They noted that aunts were most often subordinate to the mothers of the infants with whom they interacted. While admittedly very speculative, it may be that mothers gain some fitness benefits from having others care for their infants, and that subordinates may be preferred, as they are least likely to undertake any behavior that might demand an aggressive response from the mother. Again, we have no evidence of this, but the flip side of the coin is that given what we know about the importance of coalitions in primates, subordinates may make themselves more "profitable" partners by engaging in aunting behavior.

6.7.2   Byproduct mutualism, kinship, and allomothering in vervets

Fairbanks (1990) tested two hypotheses regarding the evolution of allomothering behavior in vervet monkeys *(C. aethiops sabaeus):* (1) mothers benefit by being the recipient of such actions, and this results in reducing current parental investment, which in turn leads to a shortening of between-birth time intervals, and (2) juveniles who undertake allomothering duties increase the probability of successfully raising their first child.

In her field study, Fairbanks found that although juvenile females made up only about a quarter of the group tested, 87% of infants had a juvenile female as their primary allomother. Furthermore, despite the fact mothers spent about 40% of their time away from infants, infants were rarely left alone because of allomothering behavior. Fairbanks found a significant negative correlation between the amount of time a mother's infant was carried by an allomother, and the between-birth interval of the mother (Fig. 6.15). Hypothesis 2 was supported in that although young females (3 to 5 years) typically have lower infant survival rates than older mothers (Fairbanks and McGuire, 1984), juveniles who undertook allomothering duties were indeed more likely to have their first child survive. In addition to this evidence of byproduct mutualism, kinship also plays a role in this example, as individuals who have juvenile siblings are more likely

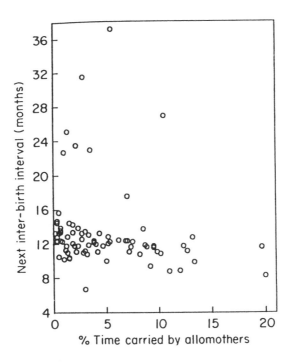

Fig. 6.15. Next interbirth interval for vervet monkey mothers as a function of the percentage of time infants were carried by allomothers (after Fairbanks, 1990).

to direct their allomothering behaviors to their sibling, rather than an unrelated group member.

6.7.3 Other examples of alloparenting

Packer et al. (1992) found that nonoffspring nursing is more common in monotocous (one offspring per litter) than polytocous (greater than one offspring per litter) mammals. Despite the fact that most primates are monotocous, and despite the enormous costs of lactation in many mammals (Millar, 1977; Gittleman and Thompson, 1988), nonoffspring nursing is common in primates (see Packer et al., 1992, for a comparative analysis of nonoffspring nursing in 100 species of mammals). For other examples of primate alloparenting behavior the reader is directed to Hrdy (1976), McKenna (1979, 1981), Altmann (1980), Reidman (1982), Nicolson (1987), Thierry and Anderson (1986), and Solomon and French (1996).

Some alloparenting behaviors in primates fall into a "gray zone" with respect to cooperation. For example, male baboons will often carry infants with them during interactions with other male group members (e.g., Deag and Crook, 1971; Altmann, 1980; Packer, 1980; Taub, 1980; Busse and Hamilton, 1981; Strum, 1984). Deag and Crook (1971) hypothesized that such infant carrying was "agonistic buffering" in that males carried the offspring of their "opponents" and that this diffused a situation that might normally turn aggressive and allowed the carrier access to resources that it might not otherwise have. Busse and Hamilton (1981), however, argued that males carry their own offspring, and hence the behavior is best thought of as parental care. In gelada baboons *(Theropithecus gelada),* Dunbar (1984b) found that both types of infant carrying occur, depending on the situation. Dunbar (1984b) also suggested a third function of infant carrying: it helps coalition formation between the carrier and the infant's mother (who might come to the aid of her infant should the carrier be attacked). Whether that function of carrying qualifies as cooperation, in that it is involved in coalition formation, is certainly debatable.

In the saddle-backed tamarin, *Saguinus fuscicollis,* males often carry infants that may have been sired by other males (Terborgh and Wilson-Goldizen, 1985). Marmosets and tamarins, however, are unique among primates in that most of the species in these groups give birth to twins. Terborgh and Wilson-Goldizen (1985) suggest that polyandry and infant carrying in *S. fusiccollis* exist because giving birth to twins, which together may weigh 25% of the mother's body weight, is energetically very costly, and a female and a single male cannot provide all the care needed to raise twins. As such, a second male is "permitted" by both the male and female to "share" reproductive access to the female, provided that it aid in the care of young. Males then cooperate with each other and the female by helping to raise the offspring, in exchange for future reproductive rights.

6.8 Summary

Given what we know about human behavior, perhaps we should not be surprised at the frequency with which primate cooperation involves reciprocity and/or some sort of coalition formation. No doubt, part of the reason that reciprocity and coalitions appear to be so common in primates is that we are more likely to search for them here. The cognitive abilities of primates are well documented, and so in many ways it makes sense to search for what might be construed as the most cognitively complex type of cooperation—reciprocity. Reciprocity, however, does not capture the entire picture, as byproduct mutualism, group selection, and, of course, kin selection also play a role in primate cooperation. In this chapter, I have reviewed examples of these four categories of cooperation in the context of social grooming, coalition formation, hunting, intergroup conflict, anti-predator behavior, and alloparenting.

7

Cooperation in eusocial insects

The ants and termites have renounced the "Hobbesian war," and they
are the better for it. Their wonderful nests, their buildings, superior in
size relative to man; their paved roads and overground vaulted galleries;
their spacious halls and granaries; their cornfields, harvesting and "malt-
ing" grain; their rational method of nursing eggs and larvae, and of
building special nests for rearing aphids whom Linneaus so pictur-
esquely describes as "the cows of the ants" and finally, their courage,
pluck and superior intelligence—all these are the normal outcome of
the mutual aid which they practice at every stage of their busy and
laborious lives. (Kropotkin, 1908, p. 37)

7.1   Introduction

At the outset, I had intended this chapter to be titled "Cooperation in Insects."
I quickly realized, however, that this was probably a more appropriate title for
a *series* of books, rather than a book chapter. At that point, "Cooperation in
Insects" became "Cooperation in Social Insects," until it became apparent that
even that was much too broad an area to even begin to review and compress
into a single chapter. And so, we have "Cooperation in Eusocial Insects." Even
after such severe pruning, however, it is fair to say that I am bound to fail at
reviewing this topic as well, and at best can hope to touch on the tip of the
iceberg. So be it. The empirical and theoretical literature on cooperation in
eusociality is so massive that if I just convey the essence of cooperation in this
group and tie it to existing theory, I will feel justified in my attempt.

Any chapter on cooperation in eusocial insects is obliged to begin with an
overview of eusociality—the phenomenon that has made the study of social
insects so alluring to evolutionary and behavioral biologists over the years.
Although the current definition of eusociality is a matter of some debate (Gadagkar, 1994; Sherman et al., 1995; Crespi and Yanega, 1995; Keller, 1995b),
the term "eusocial," as originally proposed by Batra (1966), refers to species in
which the founder of a nest survives to interact with her offspring, and in which
a division of labor exists. Michener (1969, 1974) added a third characteristic to
those of Batra (1966), namely, recognizable castes (workers and egg layers),

and Wilson (1971) and Holldobler and Wilson (1990) broadened the definition of eusociality to those species which display

> cooperation in caring for the young; reproductive division of labor, with more or less sterile individuals working on behalf of individuals engaged in reproduction; and overlap of at least two generations of life stages capable of contributing to colony labor. (p. 638)

There are many reasons why evolutionary biologists and behavioral ecologists are fascinated with eusocial insects. Their intricate social organization cannot help but remind us of our own, and that alone likely accounts for some of the interest. Whatever the other reasons behind the fascination with eusociality, a number of different theories have been proposed to explain this phenomenon. Below, I will briefly review six such hypotheses, namely, the inclusive fitness, semisocialism/mutualism, protected invasion, head start, assured fitness returns, and parental manipulation hypotheses.

7.2 The evolution of eusociality

7.2.1 Inclusive fitness theory and eusociality

The most famous of all explanations for the evolution of eusociality in insects is Hamilton's inclusive fitness theory (1963, 1964a,b, 1972: see Chapter 2 in this book for more on inclusive theory models). Hamilton, whose work was based on ideas first proposed by Fisher (1930) and Haldane (1955), summarized his model as follows:

> *The social behavior of a species evolves in such a way that in each distinct behavior-evoking situation the individual will seem to value his neighbour's fitness against his own according to the coefficients of relatedness appropriate to that situation.* (Hamilton, 1964, p. 19, emphasis in original)

Hamilton provided detailed mathematical models to support this conjecture—namely, that individuals are most likely to be cooperative and altruistic when interacting with relatives, because such relatives are more likely than average to share genes that are identical by descent. In haplodiploid species, males are haploid and females are diploid, and because of the genetics of haplodiploidy, females are related to one another on average by a coefficient of relatedness of 0.75, which has the effect of making females more related to sisters than to their own offspring. This would seem to be an ideal place to search for cooperation, particularly among females—and lo and behold we find eusociality! The effect of Hamilton's work on the study of insect social behavior (and evolutionary approaches to behavior in general) cannot be overstated. In addition to everything else, this paper also spurred on the entire (constantly growing) literature on kin recognition (e.g., Hepper, 1991).

As Reeve (1993) notes, however, there are two problems associated with "the three-quarters relatedness" hypothesis (West-Eberhard, 1975). First, in order for this hypothesis to work, colonies must exhibit a female-biased sex ratio.

While a number of good theories have been put forth on how such a female-biased sex ratio might evolve and catalyze the kinship effect (Trivers and Hare, 1976; Godfray and Grafen, 1988; Seger, 1983), much remains to be known before we can speculate on their applicability. Second, the high coefficient of relatedness among hymenopteran sisters breaks down quickly if the colony's queen is involved in multiple matings (Alexander and Sherman, 1977; Page, 1986; Ross, 1986) or if more than one queen exists in a given colony (Holldobler and Wilson, 1990; Keller, 1993, 1995a). A good deal of empirical evidence now supports the claim that in many hymenopteran species, sisters are no more related than are mother and offspring (i.e., a coefficient of relatedness of 0.5; Queller et al., 1988; Gadagkar, 1990b, 1991b; Holldobler and Wilson, 1990; Ross and Carpenter, 1991), and that hymenopterans are in possession of anywhere from crude to acute kin recognition abilities (see Gadagkar, 1985a; Gamboa et al., 1986; Breed and Bennett, 1987; Michener and Smith, 1987; Page and Breed, 1987; Carlin, 1988, 1989; Gamboa, 1988; Greenberg, 1988, for reviews of kin recognition in insects). It is important, however, to recognize that although this evidence weakens the three-quarters relatedness hypothesis, it in no way suggests that kinship per se is unimportant in the evolution of eusociality.

7.2.2 The protected invasion hypothesis and eusociality

Reeve (1993) has recently advanced the "protected invasion theory" to explain the evolution of eusociality and the lack of male parental care in the hymenoptera. This theory is put forth as a distinct alternative to kin selection and other explanations for the evolution of sociality. The idea is, in fact, elegantly simple (although the mathematics do not give that impression) and is summed up by Reeve as follows:

> Dominant alleles for maternal care in finite haplodiploid populations are more resistant to loss from genetic drift than are paternal-care alleles in haplodiploid populations or than are either maternal or paternal care in diploid populations. Similarly, alleles for female alloparental care in finite haplodiploid populations are more resistant to loss from genetic drift than are male allopaternal alleles in haplodiploid populations or than are either male or female allopaternal alleles in diploid populations. . . . Thus the protected invasion theory immediately explains all of the peculiar social features of the haplodiploid hymenoptera, namely (i) the overwhelmingly greater tendency for maternal care than paternal care in the Hymenoptera; (ii) the greater propensity for eusociality (alloparental sibling care) in Hymenoptera than in diploid species; and (iii) the greater likelihood for females than males to become alloparents (workers) in the Hymenoptera. (p. 335)

In other words, genes for "helping" are more likely to be lost in haploid males than in diploid females, which in the extreme can lead to the forms of cooperation seen among female social insects.

The protected invasion theory has recently been modified to examine the evolution of alloparental care in a wide array of species (Reeve and Shellman-

Reeve, 1995), and to examine the evolution of cooperation among unrelated individuals (Reeve and Dugatkin, ms.).

7.2.3 The head-start hypothesis and eusociality

Queller (1989) has argued that whatever role relatedness has played in the *origination* of eusociality in haplodiploid species, it may not be what *maintains eusociality*, particularly in polistine wasps. Queller decomposes inclusive fitness into the following formula:

$$r \cdot s \cdot b > r^* . s^* . b^* \qquad (7.1)$$

where r = the coefficient of relatedness between mother and offspring, s = the probability of a solitary female surviving until the time she rears her first brood to independence, and b = the number of adult offspring that survive due to the solitary's effort; r^* represents the relatedness between a worker and the young she helps, s^* is the analogous value for a worker that stays on her nest and helps, and b^* represents what the worker adds to colony productivity.

When the right hand side of the equation is greater, females should start their own nests; when it is less, they should remain on their natal nest. Queller proposes that b will often be equal to b^*, but s^* will likely be greater than s. This second inequality is what Queller refers to as "the reproductive head start of workers." That is, the probability of survival until a brood is raised to maturity is likely much greater on the natal nest than it is for females that start their own nests. This is partly because of the dangers associated with nest founding but, more to the point, because females can *immediately* start helping raise their sisters shortly after they are born if they remain on their natal nest (the "head start"). If, however, females opt to start their own nest rather than remain at home and help raise siblings, the period from birth to the point where they are able to raise their own offspring is considerably longer than the analogous period associated with staying home and starting to help. Queller uses data from *Polistes exclamans* to show that for reasonable values of the model's parameters, females should remain home and help even when the coefficient of relatedness is very low (on the order of 0.05).

In a somewhat different manner, Alexander et al. (1991) have suggested that a type of head start may have been one factor involved in the evolution of eusociality in termites and naked mole rats. They have argued that in these species, juveniles mature relatively quickly (and may be described as "mini-adults"), and are capable of helping very early on in life. Alexander et al. argue that this, plus the very safe, food-rich, expandable, multi-generation nest sites of termites and naked mole rats has made eusociality especially likely in these groups.

7.2.4 The "assured fitness return" hypothesis and eusociality

Gadagkar (1990a, 1991a) has argued that Queller's "head start" hypothesis is flawed, in that it assigns incorrect fitness values to workers. According to Gadagkar (1990a, 1991a), the head-start model gives "full credit" to a worker for raising a larvae, even though the worker may have itself come into the picture only when the larvae was near a few days shy of maturation (Fig. 7.1). This apportionment of fitness values gives workers an unfair advantage under the head-start model. As an alternative to this approach, Gadagker puts forth his "assured fitness returns" model. In this model, workers are given credit for raising a larvae *whenever* their efforts are expended, but credit is given only for the proportion of responsibility that should be assigned to a worker for its part in raising the offspring. So, for example, under the assured fitness return model, a worker who raises a larvae from birth is "credited" with more benefits than a worker who helps raise a developmentally advanced (older) larvae. What drives Gadagker's model is the "assured return" of workers, which is best understood as follows:

> A worker may care for some larvae during their very early lives and may die long before they reach the age of independence. Nevertheless she will derive some measure of fitness for her efforts because some other worker is likely to care for the same larvae and bring them to the age of indepen-

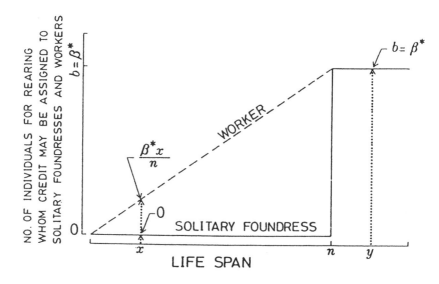

Fig. 7.1. The assured fitness return hypothesis. $b = \beta =$ the number of individuals that solitary foundresses and workers respectively can rear, if they survive the entire brood development period, and $n =$ brood development period, measured in days. Solitaries get zero credit if they survive less than n days, and b if they survive n or more days. Workers, on the other hand, get credit in proportion to the fraction of brood development that they survive with a maximum of β^* if they survive for n or more days. Thus a solitary foundress who survives x days (where $x > n$) gets zero credit, but a worker who survives for x days gets a credit of $\beta^* x/n$. A solitary foundress who survives y days (where $y < n$) gets a credit b, and a worker who survives y days gets β^*. Nobody gets more credit than b or β^* even if they survive for longer than n days because the model considers only one synchronously produced batch of brood. This is the reason for the flat portion of the curves beyond n (after Gadagker, 1990a).

dence. In contrast, a solitary foundress loses all if she dies even within a few days before her first offspring reach the age of independence. Being alone, she has no assured fitness return. (Gadagkar. 1990a, p. 19)

The assured fitness hypothesis then does not attempt to negate the role of the haplodiploid system, but rather indicates that ecological conditions may play the predominant role in shaping the evolution of eusociality, and hence hints at a non-negligible role for byproduct mutualism in the evolution of eusociality.

7.2.5   The semisocial/mutualism hypothesis and eusociality

Lin and Michener (1972), Michener and Brothers (1974), and others have argued that although kinship may be a factor in the evolution of sociality in insects, other factors such as mutualism are also likely to play a large role. Advocates of this position concentrate on

> the extent, if any, to which social behavior may arise in insects without altruism, and the allied matter of whether social behavior could originate among insects living as groups of unrelated individuals, or whether it requires for its origin groups of closely related individuals such as families. (Lin and Michener, 1972, p. 134)

Lin and Michener fall squarely on the side of the former and argue their case using a number of conceptual approaches, as well as through a selective review of the relevant literature on the Hymenopterans. One of their central points, around which this hypothesis revolves, is that the semisocial species, in which castes exist, but unrelated individuals of the same generation nest together, are more common than believed. Furthermore, they point out that transfer of individuals across colonies is not uncommon in hymenoptera, and this, in all likelihood, reduces the role of kinship. These factors, and a number of potential mutualistic benefits accrued by unrelated individuals nesting together (e.g., reduced parasite loads; Abrams and Eickwort, 1981), suggest to Lin and Michener that while relatedness may be important in *some species,* and although relatedness may act synergistically with mutualistic benefits (Queller, 1985), kinship is not necessarily involved in the evolution of eusociality in many hymenopteran species (also see Evans, 1977). Rather, Lin and Michener present what amounts to a byproduct mutualism-like model for the evolution of eusociality.

7.2.6   Parental manipulation, worker control, local mate competition, and eusociality

In his general discussion of the evolution of social behavior, Alexander (1974) defines three classes of behavior that he labels "genetically selfish in their results" (Alexander, 1974, p. 337): reciprocity, nepotism, and parental manipulation. This last class of behavior, which Alexander traces back to the work of Darwin (1859) and Fisher (1958), is described as follows:

> Parental manipulation of progeny refers to parents adjusting or manipulating their parental investment, particularly by reducing the reproduction (in-

clusive fitness) of certain progeny in the interests of increasing their own inclusive fitness via other offspring. It is easy to forget that parental care evolves, not because it increases the reproduction of individual offspring, but because it increases the reproduction of the parent. (p. 337)

Since this definition was formulated, the concept of parental manipulation has been formalized by theoreticians many times (Charnov, 1978a; Craig, 1979; Stubblefield and Charnov, 1986).

After considering the general argument for the evolution of parental manipulation, Alexander moves on to parental manipulation and the evolution of sterile castes in insects. After reviewing the case that Hamilton believed to be the best scenario for inclusive fitness to favor worker helping—a single queen mated with a single male—Alexander argues that inclusive fitness is insufficient to account for workers favoring sisters. Although it is true that in this scenario sisters are related to each other by an average r of 0.75, and to their own offspring by $r = 0.5$, Alexander raises the following issue:

> If, however, other things are indeed equal, then *queen offspring* [emphasis mine] of the above monogamous female cannot maximize their inclusive fitness by their devotion to producing offspring only half like themselves. Only if we assume that the parent has evolved to mold or manipulate her offspring phenotypically so as to maximize her own reproduction can both worker and queen offspring maximize their respective inclusive fitnesses. This they can do because of the particular phenotypes with which the mother endows each of them as a result of the distribution of parental benefits and influences. (p. 359)

Using this and similar arguments, Alexander presents the case that the parental manipulation hypothesis is better equipped to explain many of the behaviors displayed by social insects. For example:

> A principle difference between kin selection and parental manipulation is that kin selection, as formulated by Hamilton (1964a,b) requires that each individual secure genetic returns for its altruism greater than the cost of the altruism to its own personal reproduction, this return deriving from the likelihood that given relatives will carry a gene for altruism carried by the altruist. To the extent that the evolution of parental care has placed parents in the position of being able to use their investment in some offspring to increase their total reproduction via other offspring, this requirement is nullified. (p. 360)

While the parental manipulation model seems to explain some phenomena that kin selection models do not, Seger (1991) has raised an interesting problem which the parental manipulation hypothesis fails to address:

> *How* queens suppress worker reproduction is not understood for any species except the honeybee, either in a proximate (mechanistic) or an ultimate (evolutionary) sense, especially for species with large colonies, where direct intimidation of workers (as occurs in many primitively social species) would seem impossible. Various kinds of evidence suggest that pheromones are often involved (Fletcher and Ross, 1985: (Holldobler and Wilson,

1990), but why workers should be inhibited by a mere "signal" is still somewhat mysterious. (p. 356)

In contrast to parental manipulation, Trivers and Hare (1976) argue that it is the workers that control the sex ratio, not the queen. That is, they argue that *if* workers can discriminate the sex of offspring, they should invest their efforts in a 3 : 1 (female : male) ratio (this idea has been modeled many times since Trivers and Hare, 1976—e.g., see Oster et al., 1977; Benford, 1978; Charnov, 1978b; Macnair, 1978; Craig, 1980a,b; Uyenoyama and Bengtsson, 1981; Pamilo, 1982; Bulmer, 1983). But if the queen is in control of the operational sex ratio of her offspring, she should prefer a 1 : 1 ratio, as she is equally related to her male and female offspring. Trivers and Hare (1976) suggest that overall, the data available, in 1976 at least, fit the 3 : 1 hypothesis quite well (particularly in ants) and that when the data deviates from this ratio, it does so in a way that is consistent with the "worker control" hypothesis.

To complicate the "who controls the sex ratio" question even more, Alexander and Sherman (1977) argue that some of the assumptions and methods used in Trivers and Hare (1976) are flawed, and the 3 : 1 ratio uncovered in social insects is best understood not by worker control of the sex ratio, but rather as a consequence of "local mate competition" (Hamilton, 1967), wherein such female-biased numbers are a result of matings between relatives.

7.2.7 More on eusociality

The evolution of eusociality in insects has attracted enormous attention from both empiricists and theoreticians, and I do not claim to have done anything more than scrape the surface with the above review. For more on the general question of the evolution of sociality in insects, I refer the reader to Sturtevant (1938), Williams and Williams (1957), West-Eberhard (1969, 1975, 1978), Wilson (1971), Evans (1977), Starr (1979, 1985) Jeanne (1980), Craig (1983), Seger (1983), Brockmann (1984), Fletcher and Ross (1985), Gadagkar (1985a,b, 1990b,c), Joshi and Gadagkar (1985), Hansell (1987), Bourke (1988), Kukuk et al. (1989), Strassmann and Queller (1989) and Crozier and Pamilo (1996).

As I indicated above, the literature on cooperation in eusocial insects is mammoth (pardon the pun), and I make no pretexts about covering even a small fraction of it. Rather, I have chosen a number of examples that I believe are interesting, capture the essence of cooperation in insects, and help us understand the various factors that may contribute to the evolution of cooperation in insects.

7.3   Worker "policing" in honeybees

The issue of whether workers in eusocial insect species lay eggs, and what happens to such eggs, has been a contentious one for a hundred years (Seger, 1989a). Using inclusive fitness inspired thinking, Ratnieks and Visscher (1989) realized that in honeybee (*Apis mellifera*) colonies in which a queen mates with a single male, workers are more related to their nephews than to their brothers (average $r = 0.375$ versus 0.25). However, if the queen mates with multiple males, this inequality may reverse itself and workers may become more closely related to brothers than nephews (with average r depending on the number of different males with which a queen mates). Under such circumstances, Ratnieks and Visscher (1988, 1989) argue that worker "policing" (apparently named after Oster and Wilson's 1978 term "queen policing") may evolve. This policing might entail workers eating (or by some other means killing) males produced by their sisters. Ratnieks and Visscher (1989) examined this possibility using the honeybee *(A. mellifera)*, a species in which queens typically mate with some 10–20 different males.

Ratnieks and Visscher found that honeybee workers showed remarkable acumen in discriminating between worker-laid versus queen-laid eggs. After 24 hours, only 2% of the worker-laid eggs remained intact, while 61% of queen-laid eggs remained unharmed (see Fig. 7.2). Control experiments also suggest that the cue that workers use in discriminating between eggs is a queen-specific egg-marking pheromone, as predicted by Ratnieks and Visscher (1989; see Ratnieks, 1995, for more on this). Cooperation in the context of policing appears to contain both elements of kinship and nonkin-based group selection. As Ratnieks and Visscher (1989) note in the closing words of their article:

> By denying individual workers the alternative of "selfish" reproduction, in which they could increase their inclusive fitness by manipulating colony resources, worker policing aligns the genetic interests of all workers and the queen, enhancing selection for workers to increase their inclusive fitness by increasing colony resources through cooperation. (p. 797)

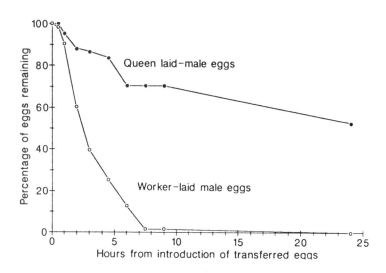

Fig. 7.2. Removal of worker- and queen-destined eggs in honeybees (after Ratneiks and Visscher, 1989).

One way to interpret honeybee policing is as follows: workers receive inclusive fitness benefits by policing in that they raise colony members to whom they are more related. However, in addition to this, there is a group productivity benefit accrued by workers, in that without policing a much greater degree of within-colony aggression would exist, and this, in turn, could decrease group productivity. This group productivity argument is separate from the issue of kinship, despite the fact that colony members are closely related, in that similar *kinds* of (group productivity related) arguments could be made even if relatedness was zero in these groups. That is, in theory, policing could occur for any number of reasons, some of which are distinct from the issue of kinship.

7.4 Honeybee colonies as superorganisms: foraging, anti-predator behavior, and hive thermoregulation

The idea of a honeybee colony and, more generally, the social insect colony as a "superorganism" is at least as old as *The Origin of Species* (1859; for general reviews see Weismann, 1893; Wheeler, 1911, 1928; Emerson, 1939; Wilson, 1971; Oster and Wilson, 1978; Lumsden, 1982; Wilson and Sober, 1989; Seeley, 1985, 1989, 1995; Holldobler and Wilson, 1990). In his discussion of the honeybee colony as a candidate for the title of superorganism, Seeley writes:

> it is perhaps not surprising that even in the most advanced insect societies, such as colonies of army ants, fungus-growing termites, or honeybees, the differentiation and integration of a society's members have not reached the point at which each member's original nature has been erased. A colony of honeybees, for instance, functions as an integrated whole and its members cannot survive on their own, yet individual honey bees are still physically independent and closely resemble in physiology and morphology the solitary bees from which they evolved. In a colony of honey bees two levels of biological organization—organism and superorganism—co-exist with equal prominence. The dual nature of such societies provides us with a special window on the evolution of biological organization, through which we can see how natural selection has taken thousands of organisms that were built for solitary life, and merged them into a superorganism. . . . It seems correct to classify a group of *organisms* as a superorganism when the organisms form a cooperative unit to propagate their genes, just as we classify a group of *cells* as an organism when the cells form a cooperative unit to propagate their genes. (1989, p. 546–548)

Although *all* members of a group rarely cooperate *fully* in any context, there is little doubt that aside from humans (and perhaps including humans; Wilson and Sober, 1994), social insect groups are the best candidates for the title of superorganism. Here, I examine the honeybee colony as a superorganism and consider this argument in the case of foraging, anti-predator behavior, and thermoregulation. If honeybee colonies operate as a superorganism, then selection between colonies—group selected behavior—may be a potent factor in the evolution of cooperation in this species.

7.4.1 Honeybee foraging

The ergonomics of social insect foraging has been examined in some detail (see Schmid-Hempel, 1991, for a recent review). For example, Seeley and his colleagues (Visscher and Seeley, 1982; Seeley, 1983, 1985, 1986, 1991, 1992, 1995; Seeley and Levien, 1987; Seeley and Towne, 1989, 1992) have done extensive work on how a honeybee colony acts as a foraging superorganism. It is to this work that I now turn.

Honeybee foraging can often involve thousands of workers and cover vast areas of ground, at least from a honeybee's perspective (Visscher and Seeley, 1982). Given that, how do individual foragers monitor their own intake rate, the changing distribution of resources through time, and the colony's needs? Although the "waggle dance" first described by von Frisch (see von Frisch, 1967, for more details on this incredible behavior, and Michelsen et al., 1991, for a recent review) provides some clue as to how workers know where a food source is, it is clearly only a partial answer to the above questions.

Lindauer (1948) demonstrated that when honeybees in a colony gather nectar at a high rate, individuals restrict their foraging to high energy food sources. *How* this colony-level response to changes in the food supply occurs remained untested until recently. Seeley and Towne (1989) addressed this issue by testing: (1) whether a colony's nutritional status was a function of storage space (empty versus filled) as well as the rate of nectar intake, (2) how foragers remained informed about the changing status of the colony's food reserves, and (3) how the mathematical theory of queues (Morse, 1958) helps explain question 2.

Working with a combination of "trained" and untrained bees, Seeley and Towne constructed artificial feeders in a natural population of honeybees in upstate New York. Clear evidence was uncovered that foragers were recruiting more nestmates to food sources when colony food resources were low and that increased recruitment was facilitated by changes in the foragers' dance patterns when arriving in the hive. The critical question is then, How did the foragers sense the change in colony-stored resources? After ruling out the possibility that foragers "patrol" the colony and get direct evidence of food reserves (as suggested by Lindauer, 1952), or that such information is gathered indirectly via the odor of empty combs (Rinderer, 1982), Seeley and Towne demonstrate that foragers gauge colony-level food supplies by using the time it takes to unload their booty to "food storer" bees as a cue (Fig. 7.3), a hypothesis first proposed by Lindauer (1948). Seeley and Towne note:

> This hypothesis also neatly explains how nectar foragers might sense a change in their colony's nutritional status due to a change in the colony's rate of nectar intake, since it is known that as a colony's intake rate increases, the food storers become busier and busier and the nectar foragers experience increasing difficulty of unloading (Lindauer, 1948; Seeley, 1986). (p. 190)

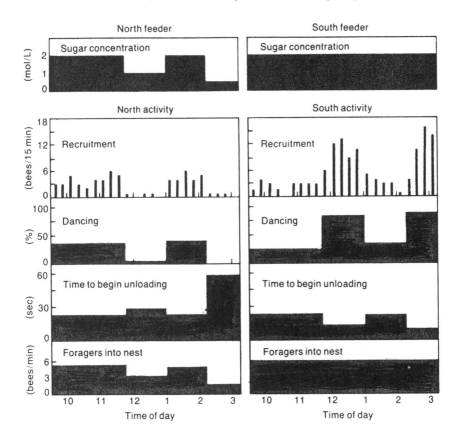

Fig. 7.3.   Honeybees respond to changes in food distribution. The quality of the north feeder was decreased, increased, and decreased again, while the quality of the south feeder remained fixed. When the food at the north source was decreased, the bees increased recruitment to the south feeder. Seeley (1989) argues that this can be explained as follows: when the north feeder declined, the bees foraging there lowered their foraging rate, which in turn lowered the number of foragers arriving at the hive. This caused foragers from the south feeder to experience shorter waiting periods when unloading nectar. This then caused foragers at the south to increase their production of recruitment signals, producing an increased number of foragers at the south feeder (after Seeley, 1989).

Last, Seeley and Towne propose that using the mathematical theory of cues—originally designed to help economists predict when to reorder merchandise—will help explain why "the time to unload" is such a good cue for foragers to use when assessing colony-level food reserves.

The literature in honeybee foraging suggests that the colony, at least with respect to foraging, operates like a well-oiled machine. The analogy of the superorganism may therefore be an appropriate one, as it is easiest to view cooperation in honeybee foraging as a result of competition between colonies in how effectively they monitor the environment and their energy reserves (for direct evidence that foraging patterns may be a colony-level trait subject to natural selection for "high" and "low" foraging lines, see Hellmich et al., 1985; Calderone and Page, 1988, 1991, 1992; Page and Fondrk, 1995; Page et al., 1995; and for more on honeybee foraging strategies in general, see the work of Schmid-Hempel and colleagues; Schmid-Hempel et al., 1985, 1987; Schmid-Hempel, 1986, 1987, 1990, 1991; Houston et al., 1988; Wolf and Schmid-Hempel, 1990).

7.4.2 Hive thermoregulation in honeybees

Hive thermoregulation in honeybees requires the collective action of many individuals, and as such it may be a legitimate candidate for a superorganismic trait. The evidence suggests that although individual honeybees are capable of thermoregulating themselves to some extent (see Heinrich, 1987, for a review), keeping the hive at a relatively constant temperature requires more than the summed action of each bee's attempt to thermoregulate itself (Southwick and Mugaas, 1971; Southwick, 1982, 1983). Heinrich (1987) suggests that group selection (in this case, hive-level selection) might account for cooperative thermoregulation in honeybees:

> Individual bees within the hive such as drones (Cahill and Lustick, 1976), could potentially be passive with regard to thermoregulation and still experience near optimal temperatures. However, if the workers all adopted the same passive strategy, then sub-optimum hive temperatures might result. (p. 105)

In other words, selection within hives may favor letting others expend the energy necessary for hive thermoregulation, but selection between hives favors a finely tuned thermoregulating superorganism. If the honeybee colony is a superorganism, then analogies to vertebrate thermoregulation may provide some testable hypotheses:

> In a homeothermic vertebrate the thermoregulatory responses are controlled in the hypothalamic region of the brain. Sensory receptors on the organism's periphery and in the hypothalamus itself monitor body temperature. When the body temperature declines below specific set-points located in the hypothalamus, neural commands are sent to the appropriate organs and these effect a coordinated response to the temperature challenge. Similarly, if a honeybee swarm or colony has bees on the periphery that act as receptors of temperature changes, which convey information about temperature to others that respond appropriately independently of their own needs then the analogy of the "superorganism" might hold. If on the other hand bees receiving the temperature information on the periphery do not communicate the information to hivemates within the core so that they can act appropriately to counteract the stimulus, then "superorganism" is inappropriate as a tool for promoting a greater understanding of the underlying mechanisms. (Heinrich, 1985, p. 396)

Heinrich (1981) constructed experiments to test if bees at the core were in fact responding to the temperature conditions of the bees at the mantle, but found no evidence that this was occurring. Based on this, and other data—primarily that *individual* bees behave in ways that keep their own body temperatures near 35° C (the hive optimum)—he rejects the superorganism analogy. Rejection of the superorganism concept here may be valid, but two caveats are in order. First, the vertebrate analogy itself may be a bad one, and the colony may be thermoregulating as a superorganism, but in a very different manner from that suggested above. Further tests are needed to address this possibility. Second, it may very well be the case that warming the hive temperature to a stable 35° C is simply a result of individuals behaving in such a way as to keep their temperature at 35° C when in the hive. Such behavior, however, is very costly in terms of energy (Heinrich, 1987), and hence if an individual could maintain such a temperature just from being in a hive, the temptation to "cheat" is certainly present. That is, the fact that individual honeybees will try and keep their body temperature at 35° C, in and of itself, is not prima facie evidence against the superorganism concept (see Southwick, 1983, for a more sympathetic view of a honeybee colony as a thermoregulating superorganism).

Warming up the hive, however, is only one type of thermoregulation. When temperatures are too high, Lindauer (1954) found that honeybees began fanning and carrying in water, which evaporated and cooled the hive. Foragers then shifted to more diluted nectar, or even water, when the situation became severe enough. Bees receiving this water regurgitated it, and had it evaporate on their mouthparts. Whether this is "superorganismic" or not is a matter of debate. Such evaporation does help the hive, but it also directly cools off the bees involved in the process (Heinrich, 1985). Similar arguments can be made about "fanning"—another response to high temperatures.

7.4.3 Honeybee colony defense

Anyone ever attacked by a nest of honeybees will probably have little difficulty grasping the concept of the honey bee colony as a superorganism. In fact, the suicidal nature of colony defense has worked its way into folklore. The more researchers examine anti-predator behavior in honeybees, the more we realize that this behavior is complex and almost mindbogglingly cooperative. Not only are there distinct "defender castes," but work on subfamilies within a hive suggests that the role of hive defender appears to have a genetic component (Robinson and Page, 1988; Breed et al., 1990; Frumhoff and Baker, 1988; Fig. 7.4). Furthermore, there are individuals who (depending on age) are more likely to be involved in nest defense than others, but even this distinction is not fine-grained enough to explain the complex cooperative anti-predator behavior displayed by honeybees (Breed et al., 1990). Colony defenders can be further subdivided into two groups: defenders, who attack very dangerous predators and guarders, who guard the nest from inter- and intraspecific parasites and may attack less dangerous threats to the nest (Moore et al., 1987; Breed et al., 1990). Lastly, ecological factors play a large role in what type of colony defense is used by different honeybee species (Seeley and Seeley, 1982).

The idea of functional organization above the level of the individual is still a contentious issue (Wilson and Sober, 1994). The superorganism concept was

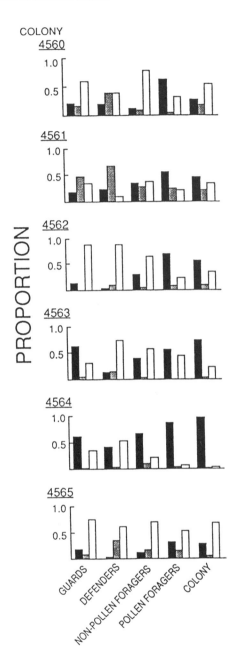

Fig. 7.4. Genotypic composition of guards, defenders, nonpollen foragers, pollen foragers, and sample of the entire colony. Colored bars represent different, electrophoretically distinguishable subfamilies within a colony (after Breed et al., 1990).

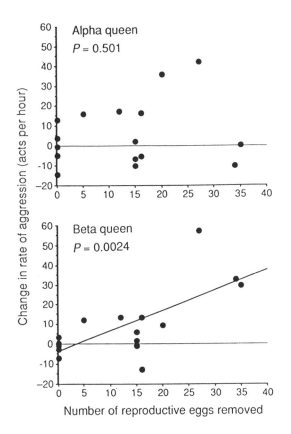

Fig. 7.5. Change in rate of aggression between control and treatment observation versus the number of eggs removed from the alpha (top panel) and beta (bottom panel) queens (after Reeve and Nonacs, 1992).

essentially abandoned in the 1960s, but has resurfaced as a bona fide subject of discussion once again. If such functional organization at the group level exists, the honeybee colony is the perfect place to look for it. Despite within-colony conflict (which certainly exists), the evidence gathered to date is at the very least consistent with the idea of the honeybee colony as a functionally organized superorganism (Seeley, 1995).

7.5 "Social contracts" in paper wasps

Theoretical work suggests that, under some circumstances, dominant individuals may tolerate the presence of subordinates on their nest, and may even allow some limited reproduction by such individuals in exchange for some type of aid on the part of subordinates (Vehrencamp, 1983; Reeve, 1991; Reeve and Ratnieks, 1993; see Keller and Reeve, 1994, for a review of these models). Reeve and Nonacs (1992), however, have argued that for such a "social contract" to be an ESS, players must have a means for punishing cheating on the part of others. How social contracts are enforced in animals is a fascinating, but perplexing question. For example, at first glance, it would appear that the punishment invoked in breaking such contracts would be much easier for dominants to enforce than for subordinates, because of the inherent asymmetries in such relationships (see Clutton-Brock and Parker, 1995, for more on punishment in animal societies).

Reeve and Nonacs (1992) examined whether subordinate female queens in the paper wasp *(Polistes fuscatus)* could in fact punish any cheating on the part of dominants. In paper wasps, dominant and subordinate queens lay worker- and queen-destined eggs, each at particular, well-established points in the nest cycle. It is critical to note that in this system, after the first workers emerge, subordinates disappear, but dominants remain at the nest (Reeve, 1991). Reeve and Nonacs hypothesized that subordinate (beta) queens should be particularly sensitive to the possibility of cheating on the part of the dominant (alpha) queen at the time when the subordinate has queen-destined eggs in the nest. In order to test this hypothesis, worker- and queen-destined eggs of both alpha and beta queens were removed in various treatments, simulating oophagy on the part of the female whose eggs were not removed. Results clearly indicate that the alpha individual does *not* respond to the simulated oophagy of worker- or queen-destined eggs by the beta. As predicted, the beta queen does not increase her aggressiveness to simulated oophagy when the eggs are worker-destined, but greatly increases such actions when queen-destined eggs are removed from the nest (Fig. 7.5). Why, however, should an alpha queen not respond to "cheating" on the part of the beta? Reeve and Nonacs (1992) suggest that as opposed to the beta queen, who disappears from the nest and has just one chance to produce queen-destined eggs, alpha queens remain on the nest and have little to gain and much to lose by forcing beta queens out before they leave naturally, as this increases the probability of nest usurpation.

Reeve and Nonac's social contract experiments suggest a complex intertwining of byproduct mutualism and reciprocity. It seems that in addition to the "power" asymmetries in an alpha queen/beta queen relationship, cooperation on the part of the dominant is related to byproduct mutualism, but cooperation on the part of subordinates is driven by reciprocity. That is, the payoffs associated with both joint nesting and punishment appear to make it in the best interest of alpha queens to remain at the nest even in the face of some rather nasty actions on the part of beta. For beta, however, the costs and benefits associated with joint nesting appear to favor reciprocity, as the destruction of precious queen-destined eggs leads to swift and severe retaliation on the part of beta queens.

7.6 Colony founding

In "The number of queens: an important trait in ant evolution," Holldobler and Wilson (1977) argue that "metrosis"—the number of queens founding a nest— is a critical trait in social insect evolution. In the vernacular of the insect world, haplometrosis refers to the case of a single queen starting a nest, and pleometrosis is used when a nest is founded by more than one queen (Fig. 7.6). Here I will concentrate on pleometric colony foundation, and in particular on ants, where cofounding queens are usually unrelated (Holldobler and Wilson 1990; Strassmann, 1989). As I argue below, pleometrosis in ants may be a good example of cooperation via group selection, while pleometrosis in wasps seems to be driven by kin selection.

7.6.1 Pleometrosis in ants

Cooperative colony foundation occurs in a number of species of ants (see Nonacs, 1988; Strassmann, 1989; Holldobler and Wilson, 1990; Keller, 1991, 1995a for reviews); in all species studied, however, cooperating cofoundresses are not closely related, suggesting that selective forces other than kin selection may favor such cooperation in natural populations (Rissing and Pollock, 1988; Strassmann, 1989). Intracolony cooperation in these species appears to be the result of intercolony aggression and territoriality (Bartz and Holldobler, 1982; Tschinkel and Howard, 1983; Rissing and Pollock, 1986, 1987; Rissing et al., 1989), such that group productivity is a function of the number of cooperators in a group (Rissing and Pollock, 1986; Fig. 7.7).

7.6.2 *Messor pergandei*

Cooperative colony foundation among unrelated foundresses has been studied extensively in the desert seed harvester ant *Messor pergandei,* one of the more common animals in the Sonoran and Mohave deserts of North America. Colonies are usually initiated by multiple foundresses (Pollock and Rissing, 1984), and behavioral (Rissing and Pollock, 1986) and genetic (Hagen et al., 1988) work on *M. pergandei* indicates cofounding queens aggregate randomly with respect to relatedness. Adult colonies are very territorial (Wheeler and Rissing, 1975; Ryti and Case, 1984), and "brood raiding" is seen among young starting colonies in the laboratory. Brood captured by nearby colonies are raised within the victorious nests, and colonies which lose their brood in such interactions die (but see Tschinkel and Howard, 1992, and Nonacs, 1993, for an alternative interpretation for such behavior). All queens produce workers and there exists a positive, linear relationship between the number of cooperating foundresses in a colony and the number of initial workers (= brood raiders) produced by that colony (Rissing and Pollock, 1991). Nests with many workers are then more likely to win brood raids (Rissing and Pollock, 1987). Until workers emerge, queens do not fight and no dominance hierarchy exists. However, once the workers have been produced, queens fight to the death and only a single queen survives.

The scenario depicted above is precisely what is necessary for group selection hypotheses to be invoked. That is, group selection requires the differential productivity of groups based on some trait. In the case of *M. pergandei,* the trait of interest is the production of workers, which, although selected against within groups (via the cheater problem), may be selected for as groups with many cooperators survive brood raiding (i.e., differential productivity of groups). The interplay of intracolony cooperation and intercolony aggression in

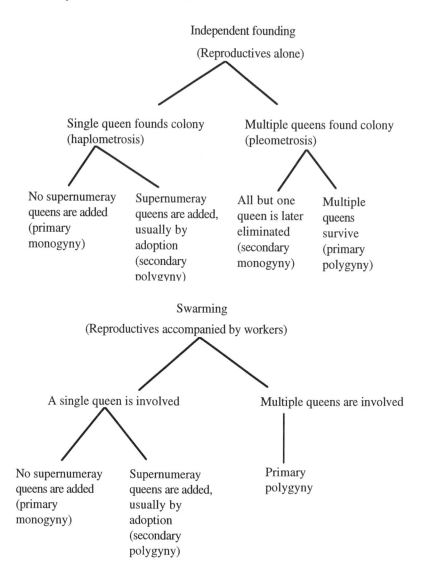

Fig. 7.6. A classification of colony founding and later colony composition with specific reference to queen number (after Holldobler and Wilson, 1979).

this ant species occurs in populations where environmental factors aggregate starting colonies, which occur only in the sandy ravine bottoms where soil moisture is available (Rissing and Pollock, 1988). Similar environmental forces causing aggregation of starting nests (as well as adult territoriality and brood raiding among starting nests) occur in all other ant species displaying cooperative colony foundation. However, a population of *M. pergandei* without cooperative colony foundation has been discovered at the Deep Canyon Biological Reserve in southern California (Ryti, 1988). In this population, foundresses are aggressive toward any other individual trying to join their nest. Work comparing these two populations (those in which pleometric queens cooperate, and those in which they do not) may shed light on the ecological underpinnings of cooperation in this species.

The idea that trait-group evolution drives cooperative colony foundation in *M. pergandei* (as argued in Dugatkin et al., 1992; Mesterton-Gibbons and Dugatkin, 1992, and Wilson, 1990) has recently been challenged. First, Nonacs (1993) has argued that the term "brood raiding" should be replaced by "nest consolidation," because in many species, both queen and workers "peacefully" abandon their nests and join with the "raiding" party. Large colonies that result from such consolidation reach advanced ergonomic capabilities more quickly that smaller ones, hence favoring such consolidation (i.e., polydomy). Second, Pfennig (1995) constructed a field experiment contrasting single and double foundress associations in *M. pergandei*. Not only did pleometric nests not outlive haplometric nests, but no brood raiding at all was observed. Both of these findings cast doubt on the efficacy of trait-group selection as a force in cooperative colony foundation. Pfennig (1995) believes that in *M. pergandei*, pleometrosis evolved because queens are forced to join *any* nest in order to minimize predation and the probability of desiccation, and that queens inhabiting a nest allow "joiners" because of the high cost of fighting.

7.6.3   *Acromyrmex versicolor*

One of the strongest cases of trait-group selected cooperation during pleometrosis is Rissing et al.'s (1989) work on *Acromymex versicolor*. *A. versicolor* nests in shady areas and shares many of the characteristics described for *M. pergandei*. Many nests are found via pleometrosis, no dominance hierarchy exists among queens, queens are unrelated, all queens produce workers, and brood raiding among starting nests appears to be common as in *M. pergandei*, with the probability of the nest surviving the brood-raiding period being a function of the number of workers produced.

A. versicolor, however, differs from *M. pergandei* in that queens in the former forage after colony foundation (*M. pergandei* do not forage outside during the "claustral" stage of early colony foundation). As a result of increased predation pressure and parasitization, foraging is a very dangerous activity for a queen, but once a queen takes on the role of forager, she remains in that role. In fact, a forager may even be punished by her nestmates if she stops assuming her food-gathering role (Rissing, personal communications). How a *single* queen becomes the group's forager is still not well understood. It does appear, however, that this decision is not a coercive one—that is, it is not forced upon a particular queen by other group mates (Rissing et al., 1989).

After a queen becomes the sole forager for her nest, she shares all the food brought into her nest with her cofounders—that is, the forager assumes the risks of foraging and obtains the benefits, while the other queens simply obtain the benefits, but pay no costs Once again, within-group cooperation (in this case extreme cooperation on the part of the forager) appears to lead to more workers, which in turn affects the probability that a given nest will be the one

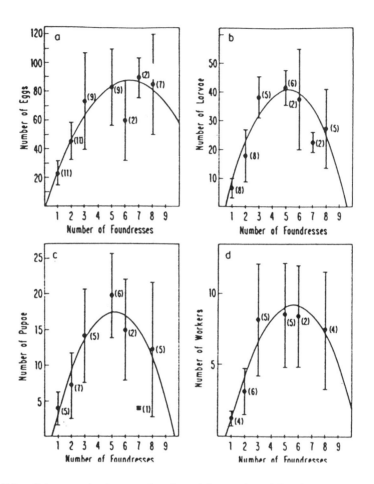

Fig. 7.7. Colony production as a function of the number of foundresses in *Myrmeco-cystus mimicus:* (a) maximum number of eggs, (b) maximum number of larvae, (c) maximum number of pupae, (d) number of workers seven weeks after of colony founding (after Bartz and Holldobler, 1982).

to survive the period of brood raiding (Seger, 1989b; Rissing et al., 1989). (See Tsuji, 1995a,b, for more on cooperation and trait-group selection in the queenless ant *Pristomyrmex pungens,* in which large workers had higher within-colony fitness, but reduced colony survival).

7.6.4 A comparison of pleometrosis in ants and wasps

In contrast to ants, in social wasps, cooperating cofoundresses are usually, if not always, closely related (West-Eberhard, 1969; Jeanne, 1972; Litte, 1977; Klahn, 1979; Michener and Smith, 1987; Queller and Strassman, 1989), and the role of kinship in pleometric cooperative colony foundation has been studied in a variety of wasp species by Queller and Strassmann and their colleagues (for more on this see Strassmann, 1981, 1989; Queller and Strassman, 1988, 1989; Queller et al., 1988, 1992, 1993; Queller, 1989; Strassmann et al., 1989, 1991, 1992; Hughes et al., 1993). Why the striking difference between ants and wasps? Why does relatedness play a critical role in the latter, but apparently no role in the former? Strassman addresses this by listing three conditions necessary to select queen-queen cooperation among unrelated foundresses:

> The first condition is that it should be hard to predict the eventual queen. . . . The second condition is that these groups produce many more brood than solitary queens. . . . The third condition is that relatives are not readily available. (Strassmann, 1989, p. 373)

Ants apparently meet these criteria far more readily than wasps.

7.7 Summary

Eusocial insects have held a special appeal to evolutionary biologists at least since the time of Darwin. The scope of cooperation in this taxa is truly magnificent, and its breadth is too great to even begin summarizing in a single chapter (or, for that matter, a single book!). At least six theories for the evolution of eusociality are currently prominent: the inclusive fitness hypothesis, the semisocialism/mutualism hypothesis, the protected invasion hypothesis, the head-start hypothesis, the assured fitness returns hypothesis, and the parental manipulation hypotheses, and these theories were reviewed in this chapter. In addition, in order to get a flavor of cooperation in this amazing taxa, the following cases of cooperation were examined in the light of the cooperator's dilemma and other modern theory—worker "policing" foraging, anti-predator behavior, and hive thermoregulation in honeybees, "social contracts" in paper wasps, and colony founding in ants and wasps.

8

To the future

"Cheshire Puss," she [Alice] began, "would you tell me please which way I ought to walk from here?"

"That depends a good deal on where you want to go from here," said the cat. (Lewis Carroll, *Alice in Wonderland*)

There is a very funny Monty Python sketch in which John Cleese portrays a famous British writer. When reminiscing, he notes that his old mentor was fond of giving the following advice to would-be writers: "The words are all out there, now we just need to get them in the right order." In some way, I think that advice applies to the study of cooperative behavior. While not ubiquitous, cooperation is certainly widespread in the animal kingdom and sooner or later we will "get the words in the right order" and come up with a solid fundamental understanding of the evolution of cooperation. But we have a long way to go. Despite interest that can be traced back to at least the time of Darwin and although hundreds of models have been constructed and empirical work has begun testing some of these models, we have only just begun a journey that may bring us, at the end, to a complete picture of how cooperation evolves.

It may not be the most strategically sound move on my part, but I must admit that I failed to accomplish the grand vision that I originally held for this book. I imagine all book authors feel this way at one time or another. While most of it is probably my fault, I take some conciliation in that at least part of the reason I didn't completely succeed in reaching my goals was out of my control. That is, the data needed to address all the questions I dreamed of addressing are, as of yet, not available. I hope, however, that this book will spur others into tackling some of the many unsolved riddles surrounding the evolution of cooperation. The remainder of this chapter will be devoted to outlining some of the ideas I had initially hoped to address in the heart of the text, but, in the end, was forced to relegate to this closing chapter.

8.1 A key to categories of cooperation

When the notion of this book was first conceived, I had hoped to create the equivalent of a species key for categories of cooperation. The idea was that if one had enough data on various ecological and behavioral parameters, one could key out which category of cooperation was at work in the system being studied, in the sense that morphological and anatomical traits are used to key out species. Reviewers of the proposal that led to this book were, on the whole, very positive, but were virtually unanimous in their opinion that creating such a key was a pipe dream. They were right. Looking back with 20/20 hindsight, the obvious reason that this endeavor was doomed to fail was that even if the data were available—and it isn't—the cooperator's dilemma game, in its current state, is simply not amenable to such an analysis.

The cooperator's dilemma game, while capturing the essence of byproduct mutualism, group selection, reciprocity, and kinship in a single game (a major conceptual advance) is not constructed to allow one the precision to say that a particular case of cooperative behavior contains elements of, for example, both byproduct mutualism and reciprocity. Although some cases of cooperation fall neatly into one category, many examples will no doubt be more difficult to pigeonhole, and so this is not a trivial problem. There is no reason in principle, however, that the theory could not be modified to handle such cases. Modifications that will make this possible may not be easy, but my colleague, Dr. Michael Mesterton-Gibbons, and I are currently working on the problem.

8.2   Quantifying the payoff matrix

Even if a family of models were constructed that could do all that was described above, we would still face the enormous challenge of quantifying the payoff matrix involved in any game. This has proven much more difficult than originally expected. In fact, the lack of data on payoff matrix entries is often the first and strongest critique laid out in criticisms of cooperation experiments.

The fact remains that Clements and Stephens' (1995) work on byproduct mutualism in foraging blue jays is the only example in this entire book in which we know the exact values associated with a game. The way that Clements and Stephens (1995) achieved this, however, entailed choosing a behavior that they themselves admit may have never been under any specific selective pressure and then testing subjects in a hyper-controlled laboratory setting. Fair enough—I do not mean to criticize this experimental protocol, as to their credit, Clements and Stephens address the pluses and minuses of such an approach. It remains, however, that the *only* study to quantify a payoff matrix was forced to use such a technique. It may be that in the end, laboratory work may be the best, if not only, way to actually control the precise entries in a payoff matrix. If this proves to be the case, however, we can still choose to examine specific cooperative behaviors that have been under clear selection pressure, even if we choose to test them in the lab.

One could certainly make the argument that it is a sad state of affairs that more than 15 years after Axelrod and Hamilton's article on the evolution of cooperation, only a single study has managed to quantify the entries into a payoff matrix. Although this is a valid point, some caveats are in order. To begin with, behavioral ecologists, despite making great leaps since their subdiscipline was created (Gross, 1994), have not measured the costs and benefits associated with a huge majority of the behaviors they study, and so cooperation is not an aberration in this regard. Such a retort, however, only goes so far, since the statement "Everyone is as bad as I am" has no predictive power whatsoever. A stronger reply would begin by asking, Does it really matter that no one knows the exact entries in the payoff matrix? The answer to this may sometimes be "No." That is, for a game to qualify as a prisoner's dilemma, or a cooperator's dilemma, one only needs to know whether a set of *inequalities* are met. If it is possible to use indirect evidence and logic to argue that the inequalities of the game are met, then detailed quantification of the payoff matrix may not be necessary. Most studies, however, have trouble doing even this. Furthermore, knowledge of a set of inequalities may tell you if a game qualifies as a specific game, but in order to analyze what the equilibrium frequency of various strategies might be in such a game, one must know the exact payoffs. As such, whenever it is possible, quantifying payoff matrices can only better help us understand the evolution of cooperation.

8.3   Correlates of cooperation

With the advent of modern comparative techniques (Harvey and Pagel, 1991), it is now possible, at least in theory, to examine correlates of cooperation across animal taxa. If the data were available (and to date they aren't), two comparisons could be made: (1) a comparison of species (or whatever the appropriate taxonomic unit might be) in which *any* category of cooperation was evident versus those in which *none* were, and (2) a comparison of species displaying one type of cooperation versus those in which another category was uncovered.

When comparing groups (in the sense of taxonomic entities), comparative techniques could be employed to look for cognitive, group-size, and "medium" effects that correlate with cooperation. To begin with, one could test whether certain cognitive abilities were commonly associated with reciprocity, byproduct mutualism, kinship, and group selection, and whether these cognitive abilities were clustered in various taxa. For example, is reciprocity more likely in large-brained animals such as mammals? While reciprocity is certainly common in this taxon, it is also found in fish, birds, insects, and so on, and so the answer to this question is not evident beforehand.

Group size may also have a profound effect on the evolution of cooperation (Mesterton-Gibbons, 1991, 1992; Mesterton-Gibbons and Childress, 1996). To examine such effects, one could plot group size against category of cooperation and look for clusters of points. Is group selection more likely in small groups? Is byproduct mutualism found most often in large groups, where individual recognition might be more difficult? Is reciprocity most common in intermediate-sized groups, where recognition is possible, but individuals have a sufficient number of others to choose from as partners (Mesterton-Gibbons, 1992b)? These and many other such questions all remain unaddressed.

Norris and Schilt (1988) have suggested that life in three-dimensional dolphin schools has selected for intelligence, and subsequently cooperative behavior. Whether or not this proves true, it is a tantalizing suggestion—does the medium an animal lives in/on (water, air, land) correlate with whether cooperation is seen, and, if it does, what category of cooperation is uncovered? We are a long way from answering such questions, but they are certainly food for thought for the next wave of cooperation experiments.

8.4 Does the skew toward reciprocity in theoretical literature reflect nature?

Theoretical work on the evolution of cooperation is *heavily* skewed toward models of reciprocity. This is true even if we ignore evolutionarily-based models developed in anthropology, psychology, economics, and mathematics, wherein virtually all models focus on reciprocity. Two questions then arise: Why the skew? and, Is it representative of nature? To answer the first of these is difficult, and amounts to speculation. But since this chapter is all about speculation, I'll state my biases. Part of the reason that the literature is so skewed may be that, despite all our attempts to avoid overlaying animal behavior with human sentiments, in the end we always do so, to one extent or another. Since cooperation is so fundamental to human behavior, and because humans are quite good "score keepers," reciprocity is the most salient form of human cooperation, and hence makes its way into the theoretical literature on this basis.

Human sentiments, however, do not explain the entire phenomena. Part of the skew in the literature may be due to the fact that reciprocity may be the most perplexing, and difficult, category of cooperation to explain. After all, in byproduct mutualism it "pays" for the individuals involved to cooperate; kinship is ingrained in the field of behavioral ecology as to need no explanation; and group selection is controversial, but easy to grasp intuitively. None of these characteristics, however, holds true for reciprocity, and the prisoner's dilemma shows just how hard it is for cooperation to evolve under certain conditions. If this portrait is correct, then the skew toward reciprocity in the theoretical literature is more understandable. After all, hard problems require more work, and, besides that, there is nothing that a theoretician likes more than a challenge. All that being said, however, it is critical that more models addressing at least byproduct mutualism and group-selected cooperation (there may be enough kin-based cooperation papers out there, but even that is arguable) be developed. It has often proven to be the case that behaviors that appear easy to explain without models turn out to be more complex (and interesting) when subject to rigorous mathematical treatment. In addition, as argued above, without more models of all categories of cooperation, it will be difficult to classify a behavior as falling into any one (or more) category.

The second question—does the skew toward reciprocity in the theoretical literature represent what is happening in nature?—is also difficult to answer. My guess is that, in the long run, the answer will probably be a resounding "No." To date, most empirical studies have been constructed to examine reciprocity, but this is in all likelihood a case of empiricists following the lead of theoreticians, rather than anything else. The more controlled studies we do, the more we see that reciprocity explains only some of the cooperation found in nature. For example, recent empirical work—for example, Heinsohn and Packer's (1995) work on lions—cries out for new models that examine the interaction of various categories of cooperation and how they produce such strategies as "conditional cooperator," "unconditional cooperator," "conditional laggard" and "unconditional laggard." My guess is that the more empirical work we do, the more such complex types of cooperative systems we will uncover.

8.5 Spatial games and the evolution of cooperation

All of the original work on the evolution of cooperation essentially ignored spatial variation and more generally spatial structure in populations (this charge cannot, however, be leveled against early group-selection models of altruism). Recently, theoreticians have added spatial structure to their models of cooperation and come up with some fascinating preliminary results. The most interesting of these findings is that, even when individuals are trapped in a prisoner's dilemma (i.e., the worst-case scenario for the evolution of cooperation), cooperation can evolve *without reciprocity* under certain population structures (Nowak and May 1992; Wilson et al., 1992; Nowak and Sigmund 1993b). This may be a particularly important finding given the fact that future work on cooperation will no doubt include studies on very simple organisms and even molecules (I thank Bob May for pointing this out). Clearly, this line of inquiry needs to be pursued in future work.

8.6 The bright side and the dark side

In closing, I think it is fair to say that when writing and/or reading a book on cooperative behavior, at first glance it is hard not to come away with a picture of the world as a rather nice to place to live. Skirting the philosophical aspects of this point, from an empirical perspective this is only partly true. When rethinking the topics covered in this book, we see that cooperation often takes place in the context of rather nasty situations; that is, competition, aggression, predation, and so on, certainly play a large role in understanding cooperative behavior. So the "dark side" also reveals itself. It may be that unraveling the mystery of cooperative behavior may be the key to a comprehensive understanding of animal behavior, both nasty and nice. If this book moves us, even a little, in that direction, then the effort will have not have been in vain.

References

Abdel-malek, S. A. (1963). Diurnal rhythm of feeding of the three-spine stickleback of Kandalaksha Bay of the White Sea. *Vopr Ikhtiol (in Russian)*, **3**, 326–35.

Abel, E. F. (1971). Zur Ethologie von Putzsymbiosen einheimischer Susswasserfische im naturlichen Biotop. *Oecologia*, **6**, 133–51.

Abrams, J., and Eickwort, G. C. (1981). Nest switching and guarding by the communal sweat bee *Agapostemon virescens* (Hymeoptera, Halictidae). *Insect Soc.*, **28**, 105–16.

Alatalo, R., and Helle, P. (1990). Alarm calling by individual willow tits, *Parus montanus. Anim. Behav.*, **40**, 437–42.

Alevizon, W. S. (1976). Mixed schooling and its possible significance in a tropical western Atlantic parrotfish and surgeonfish. *Copeia*, **1976**, 796–98.

Alexander, R. D. (1974). The evolution of social behavior. *Ann. Rev. Ecol. Syst.*, **5**, 325–83.

Alexander, R. D. (1986). Ostracism and indirect reciprocity: the reproductive significance of humor. *Ethol. Sociobiol.*, **7**, 253–70.

Alexander, R. D., Noonan, K. M., and Crespi, B. J. (1991). The evolution of eusociality. In *The biology of the naked mole-rat* (ed. P. Sherman, J. U. M. Jarvis, and R. D. Alexander), pp. 3–44, Princeton, NJ: Princeton Univ. Press.

Alexander, R. D., and Sherman, P. W. (1977). Local mate competition and parental investment in insects. *Science*, **196**, 494–500.

Allee, W. C. (1931). *Animal aggregations.* Chicago: Univ. of Chicago Press.

Allee, W. (1938). *The social life of animals.* New York: Henry Schuman. Reprint, 1958, Beacon Books.

Allee, W. C. (1943). Where angels fear to tread: a contribution from general sociology to human ethics. *Science*, **97**, 517–25.

Allee, W. C. (1951). *Cooperation among animals.* New York: Henry Schuman.

Altmann, J. (1980). *Baboon mothers and infants.* Cambridge, MA: Harvard Univ. Press.

Altmann, J., Hausfater, G., and Altmann, S. A. (1988). Determinants of reproductive success in Savannah baboons. In *Reproductive success* (ed. T. Clutton-Brock), pp. 403–18, Chicago: Univ. of Chicago Press.

Altmann, S. (1979). Baboon progressions. Order or chaos? A study of one-dimensional group geometry. *Anim. Behav.*, **27**, 46–80.

Alverdes, F. (1927). *Social life in the animal world.* London: Kegan Paul, Trench and Trubner.

Aoki, K. (1983). A quantitative genetic model of reciprocal altruism: a condition for kin or group selection to prevail. *Proc. Natl. Acad. Sci. USA*, **80**, 4065–68.

Aoki, K. (1984). Evolution of alliance in primates: a population genetic model. *J. Ethol.*, **2**, 55–61.

Aristotle (328 B.C.). *Politics.* Reprint, 1947, Walter J. Black.

Axelrod, R. (1980a). Effective choices in the Prisoner's Dilemma. *J. Conf. Res.*, **24**, 3–25.

Axelrod, R. (1980b). More effective choices in the Prisoner's Dilemma. *J. Conflict Res.*, **24**, 379–403.

Axelrod, R. (1984). *The evolution of cooperation.* New York: Basic Books.

Axelrod, R. (1986). An evolutionary approach to norms. *Amer. Pol. Sci. Rev.*, **80**, 1101–11.

Axelrod, R. (1987). The evolution of strategies in the iterated prisoner's dilemma. In *Genetic algorithms and simulated annealing* (ed. L. Davis), pp. 32–41, Morgan Kaufmann.

Axelrod, R., and D'Ambrosio, L. (1994). *Annotated bibliography on the evolution of cooperation.* Institute of Public Policy paper, University of Michigan.

Axelrod, R., and Dion, D. (1987). *Annotated bibliography on the evolution of cooperation.* Inst. of Public Policy paper, University of Michigan.

Axelrod, R., and Hamilton, W. D. (1981). The evolution of cooperation. *Science*, **211**, 1390–96.

Banks, E. (1985). Warder Clyde Allee and the Chicago school of animal behavior. *J. Hist. Behav. Sci.*, **21**, 345–53.

Barash, D. P. (1975). Marmot alarm calling and the question of altruistic behavior. *Am. Midl. Nat.*, **94**, 468–70.

Barash, D. P. (1976). Social behavior and individual differences in free-living alpine marmots (*Marmota marmota*). *Anim. Behav.*, **24**, 27–35.

Barker, E. (1962). *Social contract: essays by Locke, Hume and Rousseau.* New York: Oxford Univ. Press.

Barlow, G. W. (1974). Extraspecific imposition of social grouping among surgeonfishes (Pisces: Acanthuridae). *J. Zool. Soc. Lond.*, **174**, 333–40.

Barlow, G. W. (1975). On the sociobiology of some hermaphrodictic serranid fishes, the hamlets, in Puerto Rico. *Marine Biol.* **33**, 295–300.

Barnard, C. J., and Sibly, R. M. (1981). Producers and scroungers: a general model and its application to captive flocks of house sparrows. *Anim. Behav.*, **29**, 543–50.

Barton, R. (1985). Grooming site preferences in primates and their functional implications. *Int. J. Primatology,*, **6**, 519–31.

Bartz, S., and Holldobler, B. (1982). Colony foundation in *myrmecocytus mimicus* and the evolution of foundress associations. *Behav. Ecol. Sociobiol.*, **10**, 137–47.

Batra, S. W. T. (1966). Nests and social behavior of halictine bees of India (Hymenoptera: Halictidae). *Ind. J. Entomol.*, **28**, 375–93.

Bednarz, J. C. (1988). Cooperative hunting in Harris' hawks (*Parabuteo unicinctus*). *Science*, **239**, 1525–27.

Bednarz, J. C., and Ligon, J. D. (1988). A study of the ecological basis of cooperative breeding in the Harris' hawk. *Ecology*, **69**, 1176–87.

Bendor, J., and Swistak, P. (1995). Types of evolutionary stability and the problem of cooperation. *Proc. Natl. Acad. Sci.*, USA, **92**, 3596–600.

Benford, F. A. (1978). Fisher's theory of the sex ratio applied to the social hymenoptera. *J. Theo. Biol.*, **72**, 710–27.

Bennett, G. F. (1969). *Boophilus microplus* (Acarina: ixodidae): experimental infestations on cattle restrained from grooming. *Exp. Parasit.*, **26**, 323–28.

Bent, A. C. (1938). *Life histories of North American birds.* Washington: U.S. Government Printing Office.

Bercovitch, F. (1988). Coalitions, cooperation and reproductive tactics among adult male baboons. *Anim. Behav.*, **36**, 1198–209.

Bercovitch, F. (1991). Social stratification, social strategies and reproductive success in primates. *Ethol. Sociobiol.*, **12**, 315–33.

Bernstein, I. S., and Sharpe, L. (1966). Social roles in a rhesus monkey group. *Behaviour*, **26**, 91–104.

Bertram, B. (1978). Living in groups: predators and prey. In *Behavioural ecology: an evolutionary approach* (ed. J. R. Krebs and N. Davies), pp. 64–96, Sunderland: Sinauer Assoc.

Bertram, B. (1992). *The ostrich communal nesting system.* Princeton: Princeton Univ. Press.

Bildstein, K. L. (1982). Responses of Northern harriers to mobbing passerines. *J. Field Ornithol.*, **53**, 7–14.

Bildstein, K. L. (1983). Why white-tailed deer flag their tails. *Am. Nat.*, **121**, 709–15.

Blurton-Jones, N. (1987). Tolerated theft, suggestions about the ecology of sharing, hoarding and scrounging. *Biology and Social Life*, **26**, 31–54.

Boccia, M. L. (1987). The physiology of grooming: a direct test of the tension reduction mechanism. *Am. J. Primatology*, **12**, 330.

Boehm, C. (1992). Segmentary warfare and management of conflict: a comparison of East African chimpanzees and patrilineal-patrilocal humans. In *Coalitions and alliances in humans and other animals* (ed. A. Harcourt and F. B. M. de Waal), pp. 137–73, Oxford: Oxford Univ. Press.

Boesch, C. (1994a). Chimpanzees–red colobus: a predator–prey system. *Anim. Behav.*, **47**, 1135–48.

Boesch, C. (1994b). Cooperative hunting in wild chimpanzees. *Anim. Behav.*, **48**, 653–67.

Boesch, C. (1994c). Hunting strategies of Gombe and Tai chimpanzees. In *Chimpanzee culture* (ed. R. Wrangham, W. C. McGrew, F. de Waal, and P. Heltne), pp. 77–91. Cambridge: Harvard Univ. Press.

Boesch, C., and Boesch, H. (1989). Hunting behavior of wild chimpanzees in the Tai National Park. *Am. J. Phys. Anthrol.*, **78**, 547–73.

Boorman, S., and Levitt, P. (1973a). A frequency-dependent natural selection model for the evolution of social cooperation networks. *Proc. Natl. Acad. Sci. USA*, **70**, 187–89.

Boorman, S. A., and Levitt, P. R. (1973b). Group selection on the boundary of a stable population. *Theo. Pop. Biol.*, **4**, 85–128.

Borstnik, B. B., Pumpernik, D., Hofacker, I. L., and Hofacker, G. L. (1990). An ESS analysis for ensembles of Prisoner's Dilemma strategies. *J. Theo. Biol.*, **142**, 195–220.

Bossema, I., and Benus, R. F. (1985). Territorial defence and intrapair cooperation in the carrion crow (*Corvus corone*). *Behav. Ecol. Sociobiol.*, 16, 99–104.

Bourke, A. F. G. (1988). Worker reproduction in the higher eusocial hymenoptera. *Q. Rev. Biol.*, **63**, 291–311.

Boyd, R. (1988). Is the repeated Prisoner's Dilemma a good model of reciprocal altruism? *Ethol. Sociobiol.*, **9**, 211–22.

Boyd, R. (1989). Mistakes allow evolutionary stability in the repeated Prisoner's Dilemma game. *J. Theo. Biol.*, **136**, 47–56.

Boyd, R. (1992). The evolution of reciprocity when conditions vary. In *Coalitions and alliances in humans and other animals* (ed. A. Harcourt and F. de Waal), pp. 473–89, Oxford: Oxford Univ. Press.

Boyd, R., and Lorberbaum, J. (1987). No pure strategy is evolutionarily stable in the repeated Prisoner's Dilemma. *Nature*, **327**, 58–59.

Boyd, R., and Richerson, P. (1982). Cultural transmission and the evolution of cooperative behavior. *Human Ecol.*, **10**, 325–51.

Boyd, R., and Richerson, P. (1988). The evolution of reciprocity in sizable groups. *J. Theo. Biol.*, **132**, 337–56.

Boyd, R., and Richerson, P. J. (1989). The evolution of indirect reciprocity. *Social Networks*, **11**, 213–36.

Boyd, R., and Richerson, P. (1990a). Group selection among alternative evolutionarily stable strategies. *J. Theo. Biol.*, **145**, 331–42.

Boyd, R., and Richerson, P. J. (1990b). Culture and cooperation. In *Beyond self-interest* (ed. J. J. Mansbridge), pp. 111–32, Chicago: Univ. of Chicago Press.

Boyd, R., and Richerson, P. (1992). Punishment allows the evolution of cooperation (or anything else) in sizable groups. *Ethol. Sociobiol.*, **13**, 171–95.

Breed, M. D., and Bennett, B. (1987). Kin recognition in highly eusocial insects. In *Kin recognition in animals* (ed. D. C. Fletcher and C. D. Michener), pp. 243–85, Chichester: Wiley.

Breed, M. D., Robinson, G. E., and Page, R. E. (1990). Division of labor during honey bee colony defense. *Behav. Ecol. Sociobiol.*, **27**, 395–401.

Brockmann, H. J. (1984). The evolution of social behavior in insects. In *Behavioural ecology*, 2nd ed. (ed. J. R. Krebs and N. B. Davies), pp. 340–61, Sunderland, MA: Sinauer.

Brown, C. (1985). The costs and benefits of coloniality in the cliff swallow. Ph.D. dissertation. Princeton University.

Brown, C. R. (1986). Cliff swallow colonies as information centers. *Science*, **234**, 83–85.

Brown, C. (1988). Social foraging in cliff swallows: local enhancement, risk sensitivity and the avoidance of predators. *Anim. Behav.*, **36**, 780–92.

Brown, C., Brown, M. B., and Shaffer, M. L. (1991). Food-sharing signals among socially foraging cliff swallows. *Anim. Behav.*, **42**, 551–64.

Brown, D. H., and Norris, K. S. (1956). Observations of captive and wild cetaceans. *J. Mammal.*, **37**, 311–26.

Brown, J. L. (1966). Types of group selection. *Nature*, **211**, 870.

Brown, J. L. (1970). Cooperative breeding and altruistic behavior in the Mexican jay, *Aphelocoma ultramarina*. *Anim. Behav.*, **18**, 366–78.

Brown, J. L. (1974). Alternate routes to sociality in jays—with a theory for the evolution of altruism and communal breeding. *Am. Zool.* **14**, 63–80.

Brown, J. L. (1975). *The evolution of behavior*. New York: W. W. Norton.

Brown, J. L. (1982). Optimal group size in territorial animals. *J. Theo. Biol.*, **95**, 793–810.

Brown, J. L. (1983). Cooperation—a biologist's dilemma. In *Advances in the study of behaviour* (ed. J. S. Rosenblatt), pp. 1–37, New York: Academic Press.

Brown, J. L. (1987). *Helping and communal breeding in birds*. Princeton: Princeton Univ. Press.

Brown, J. L. (1994). Historical patterns in the study of avian social behavior. *The Condor*, **96**, 232–43.

Brown, J. S., Sanderson, M., and Michod, R. (1982). Evolution of social behavior by reciprocation. *J. Theo. Biol.*, **99**, 319–39.

Bryne, R., and Whiten, A. (eds.) (1988). *Machiavellian intelligence*. Oxford: Clarendon Press.

Buitron, D. (1983). Variability in the responses of black-billed magpies to natural predators. *Behaviour*, **87**, 209–36.

Bulmer, M. (1983). Sex ratio evolution in social hymenoptera under worker control and behavioral dominance. *Am. Nat.*, **121**, 899–902.

Bull, J., and Rice, W. (1991). Distinguishing mechanisms for the evolution of cooperation. *J. Theo. Biol.*, **149**, 63–74.

Busse, C., and Hamilton, W. J. (1981). Infant carrying by male chacma baboons. *Science*, **212**, 1281–83.

Busse, C. (1978). Do chimps hunt cooperatively? *Am. Nat.*, **112**, 767–70.

Byers, J. (1984). Play in ungulates. In *Play in animals and humans* (ed. P. K. Smith), pp. 43–65, Oxford: Blackwell Scientific.

Bygott, J. D. (1979). Agonistic behavior, dominance and social structure in wild chimpanzees of the Gombe National Park. In *The great apes* (ed. D. A. Hamburg and E. R. McCown), pp. 405–425. Menlo Park, CA: Benjamin/Cummings.

Bygott, J. D., Bertram, B., and Hanby, J. P. (1979). Male lions in large coalitions gain reproductive advantages. *Nature*, **282**, 839–41.

Cade, T. J. (1982). *The falcons of the world.* Ithaca, NY: Comstock/Cornell Univ. Press.

Cahill, K., and Lustick, S. (1976). Oxygen consumption and hermoregulation in *Apis mellifera* workers and drones. *Comp. Biochem. Physiol.*, **55A**, 355–57.

Caine, N. G. (1984). Visual scanning by tamarins. *Folia Primatol.*, **43**, 59–67.

Caine, N. G., Addington, R. L., and Windfelder, T. L. (1995). Factors affecting the rate of food calls given by red-bellied tamarins. *Anim. Behav.*, **59**, 53–60.

Calderone, N. W., and Page, R. P. (1988). Genetic variability in age polyethism and task specialization in the honey bee, *Apis mellifera* (Hymenoptera: Apidae). *Behav. Ecol. Sociobiol.*, **22**, 17–25.

Calderone, N. W., and Page, R. P. (1991). Evolutionary genetics of division of labor in colonies of the honey bee (*Apis mellifera*). *Am. Nat.*, **138**, 69–92.

Calderone, N. W., and Page, R. P. (1992). Effect of interactions among genetically diverse nestmates on task specialization by foraging honeybees, *Apis mellifera*. *Behav. Ecol. Sociobiol.*, **30**, 69–92.

Caldwell, M. C., Brown, H., and D. K., Caldwell. (1963). Intergeneric behavior by a captive Pacific pilot whale. *Los. Ang. City Mus. Contrib. Sci.*, **70**, 1–12.

Caporeal, L., Dawes, R., Orbell, J., and van de Kragt, A. J. C. (1989). Selfishness examined: cooperation in the abscence of egoistic motives. *Behav. Brain Sci.*, **12**, 683–739.

Caraco, T., and Brown, J. L. (1986). A game between communal breeders: when is food sharing stable. *J. Theo. Biol.*, **118**, 379–93.

Caraco, T., and Wolf, L. L. (1975). Ecological determinants of group sizes of foraging lions. *Am. Nat.*, **109**, 343–52.

Caro, T. M. (1986a). The functions of stotting in Thomson's gazelles: some tests of predictions. *Anim. Behav.*, **34**, 663–84.

Caro, T. M. (1986b). The functions of stotting: a review of the hypotheses. *Anim. Behav.*, **34**, 649–62.

Caro, T. M. (1994). Ungulate antipredator behavior: preliminary and comparative data from African bovids. *Behaviour*, **128**, 189–228.

Caro, T. M. (1995a). Pursuit-deterrence revisted. *Trends Ecol. Evol.*, **10**, 500–503.

Caro, T. M. (1995b). *Cheetahs of the Serengeti plains: group living in an asocial species.* Chicago: Univ. of Chicago.

Caro, T. M., Lombardo, L., Goldizen, A. W., and Kelly, M. (1995). Tail-flagging and other antipredator signals in white deer: new data and synthesis. *Behav. Ecol.*, **6**, 442–50.

Carlin, N. F. (1988). Species, kin and other forms of recognition in the brood discrimination behavior of ants. In *Advances in myrmecology* (ed. J. C. Trager), pp. 267–95, Leiden: Brill.

Carlin, N. F. (1989). Discrimination with and between colonies of social insects: two null hypotheses. *Neth. J. Zool.*, **39**, 86–100.

Carpenter, C. R. (1942). Sexual behavior of free-ranging rhesus monkeys, *M. mulatta*. *J. Comp. Psychol.*, **33**, 113–62.

Carroll, J. P. (1985). Brood defense by female ring-necked pheasants against northern harriers. *J. Field Ornithol.*, **56**, 283–84.

Cavalli-Sforza, L., and Feldman, M. (1978). Darwinian selection and altruism. *Theo. Pop. Biol.*, **14**, 268–80.

Chalmeau, R. (1994). Do chimpanzees cooperate in a learning task? *Primates*, **35**, 385–92.

Chapais, B. (1992). The role of alliances in social inheritance of rank among female primates. In *Coalitions and alliances in humans and other animals* (ed. A. Harcourt and F. B. M. de Waal), pp. 29–59, Oxford: Oxford Univ. Press.

Chapais, B., and Schulman, S. R. (1980). An evolutionary model of female dominance relations in primates. *J. Theo. Biol.*, **82**, 47–89.

Charlesworth, B. (1980). Models of kin selection. In *Evolution of social behavior: hypotheses and empirical tests* (ed. H. Markl), Weinheim: Springer-Verlag.

Charnov, E. (1977). An elementary treatment of the genetical theory of kin selection. *J. Theo. Biol.*, **66**, 541–50.

Charnov, E. (1978a). Evolution of eusocial behavior: offspring choice or parental manipulation. *J. Theo. Biol.*, **75**, 451–65.

Charnov, E. L. (1978b). Sex-ratio selection in eusocial hymenoptera. *Am. Nat.*, **112**, 317–26.

Charnov, E. (1982). *The theory of sex allocation*. Princeton, NJ: Princeton Univ. Press.

Charnov, E. L., and Krebs, J. R. (1975). The evolution of alarm calls: altruism or manipulation? *Am. Nat.*, **109**, 107–12.

Charnov, E. L., Orians, G. H., and Hyatt, K. (1976). The ecological implications of resource depression. *Am. Nat.*, **110**, 247–59.

Cheney, D. L. (1977). The acquisition of rank and the development of reciprocal alliances among free-ranging immature baboons. *Behav. Ecol. Sociobiol.*, **2**, 303–18.

Cheney, D. L. (1987). Interactions and relationships between groups. In *Primate societies* (ed. B. Smuts, D. L. Cheney, R. Seyfarth, R. Wrangham, and T. Struhsaker), pp. 267–81, Chicago: Univ. of Chicago Press.

Cheney, D. L. (1994). Intragroup cohesion and intergroup hostility: the relation between grooming distributions and intergroup competition among female primates. *Behav. Ecol.*, **3**, 334–45.

Cheney, D. L, and Seyfarth, R. (1977). Behavior of adult and immature male baboons during inter-group encounters. *Nature*, **269**, 404.

Cheney, D. L., and Seyfarth, R. (1980). Vocal recognition in free-ranging vervet monkeys. *Anim. Behav.*, **28**, 362–67.

Cheney, D. L., and Seyfarth, R. (1981). Selective forces affecting the predator calls of vervet monkeys. *Behav.*, **76**, 25–61.

Cheney, D. L., and Seyfarth, R. (1982). Recognition of individuals within and between groups of free-ranging vervet monkeys. *Am. Zool.*, **22**, 519–26.

Cheney, D. L., and Seyfarth, R. (1985). Vervet alarm calls: manipulation through shared information. *Behaviour*, **94**, 150–66.

Cheney, D. L., and Seyfarth, R. M. (1990). *How monkeys see the world.* Chicago: Univ. of Chicago Press.

Clark, C. (1987). The lazy, adaptable lion: a Markovian model of group foraging. *Anim. Behav.*, **35**, 361–69.

Clark, E. (1959). Functional hermaphroditism and self-fertilization in a serranid fish. *Science*, **29**, 215–16.

Clements, K. C., and Stephens, D. W. (1995). Testing models of animal cooperation: feeding bluejays cooperate mutualistically, but defect in a massively iterated Prisoner's Dilemma. *Anim. Behav.*, **50**, 527–35.

Clutton-Brock, T. H., Greenwood, P. J., and Powell, R. P. (1976). Ranks and relationships in Highland ponies and Highland cows. *Z. Tierpsychol*, **41**, 202–16.

Clutton-Brock, T. H., and Parker, G. A. (1995). Punishment in animal societies. *Nature*, **373**, 209–16.

Cohen, D., and Eshel, I. (1976). On the founder effect and the evolution of altruistic traits. *Theo. Pop. Biol.*, **10**, 276–302.

Colgan, P. (1986). The motivational basis of fish behaviour. In *The behavior of teleost fishes* (ed. T. Pitcher), pp. 23–68, Baltimore, MD: Johns Hopkins Univ. Press.

Collias, N. E., and Joos, M. (1953). The spectrographic analysis of sound signals in domestic fowl. *Behav.*, **5**, 175–88.

Colman, A. (ed.) (1982). *Cooperation and competition in humans and animals.* Berkshire, Eng.: Van Nostrand Reinhold.

Connor, R. C. (1986). Pseudoreciprocity: investing in mutualism. *Anim. Behav.*, **34**, 1652–54.

Connor, R. C. (1992). Egg-trading in simultaneous hermaphrodites: an alternative to Tit for Tat. *J. Evol. Biol.* **5**, 523–28.

Connor, R. C. (1995a). The benefits of mutualism: a conceptual framework. *Biol. Rev.*, **70**, 427–57.

Connor, R. C. (1995b). Impala allogrooming and the parcelling model of reciprocity. *Anim. Behav.*, **49**, 528–30.

Connor, R. C. (1996). Partner preferences in by-product mutualism and the case of predator inspection in fish. *Anim. Behav.*, **51**, 451–54.

Connor, R. C., and Norris, K. S. (1982). Are dolphins reciprocal altruists? *Am. Nat.*, **119**, 358–74.

Connor, R. C., Smolker, R. A., and Richards, A. F. (1992a). Dolphin alliances and coalitions. In *Coalitions and alliances in humans and other animals* (ed. A. H. Harcourt and F. B. M. de Waal), pp. 415–43, Oxford: Oxford Univ. Press.

Connor, R. C., Smolker, R. A., and Richards, A. F. (1992b). Two levels of alliance formation among male bottle-nosed dolphins. *Proc. Natl. Acad. Sci. USA*, **89**, 987–90.

Craig, J. L. (1984). Are communal pukeko caught in a Prisoner's Dilemma? *Behav. Ecol. Sociobiol.*, **14**, 147–50.

Craig, R. (1979). Parental manipulation, kin selection and the evolution of altruism. *Evol.*, **33**, 319–34.

Craig, R. (1980a). Sex investment ratios in social hymenoptera. *Am. Nat.*, **116**, 311–23.

Craig, R. (1980b). Sex ratio changes and the evolution of eusociality by kin selection: simulation and game theory studies. *J. Theo. Biol.*, **87**, 55–70.

Craig, R. (1983). Subfertility and the evolution of eusociality by kin selection. *J. Theo. Biol.*, **100**, 379–97.

Creel, S. R. (1990). How to measure inclusive fitness. *Proc. R. Soc. Lond.*, **241**, 229–31.

Creel, S. R., and Creel, N. M. (1991). Energetics, reproductive suppression and obligate communal breeding in carnivores. *Behav. Ecol. Sociobiol.*, **28**, 263–70.

Creel, S. R., Monfort, S. L., Wildt, D. E., and Waser, P. (1991). Spontaneous lactation is an adaptive result of pseudopregnancy. *Nature*, **351**, 600–62.

Creel, S. R., and Waser, P. (1991). Failure of reproductive suppression in dwarf mongooses: accident of adaptation? *Behav. Ecol.*, **2**, 7–15.

Crespi, B., and Yanega, D. (1995). The definition of eusociality. *Behav. Ecol.*, **6**, 109–15.

Crowley, P. H., Provencher, L., Sloane, S., Dugatkin, L. A., Spohn, B., Rogers, B., and Alfieri, M. (1995). Evolving cooperation: the role of individual recogition. *Biosystems*, **37**, 49–66.

Crozier, R. H., and Pamilo, P. (1996). *Evolution of social insect colonies: sex allocation and kin selection.* Oxford: Oxford Univ. Press.

Csanyi, V. (1985). Ethological analysis of predator avoidance in the paradise fish. I. Recognition and learning of predators. *Behaviour*, **92**, 227–39.

Csanyi, V., Toth, P., Altbacker, V., Doka, A., and Gervai, J. (1985). Behavioral elements of the paradise fish, II. A functional analysis. *Acta Biol. Hung.*, **36**, 115–30.

Curio, E. (1978). The adaptive significance of avian mobbing. I. Teleonomic hypotheses and predictions. *Z. Tierpsychol.*, **48**, 175–83.

Curio, E., Ernest, U., and Vieth, W., (1978a). The adaptive significance of avian mobbing. II. Cultural transmission and enemy recognition in blackbirds: effectiveness and some constraints. *Z. Tierpsychol.*, **48**, 185–202.

Curio, E., Ernest, U., and Vieth, W. (1978b). Cultural transmission of enemy recognition: one function of mobbing. *Science*, **202**, 899–901.

Curio, E., and Regelmann, K. (1985). The behavioral dynamics of great tits (*Parus major*) approaching a predator. *Z. Tierpsychol.*, **69**, 3–18.

Curio, E., and Regelmann, K. (1986). Predator harassment implies a real deadly risk: a reply to Hennessy. *Ethology*, **72**, 75–78.

Darkhov, A. A., and Panyushkin, S. N. (1988). Cleaning symbiosis among six freshwater fishes. *J. Icth.*, **28**, 161–67.

Datta, S. (1983). Relative power and the acquisition of rank. In *Primate social relationships* (ed. R. Hinde), pp. 91–103, Oxford: Blackwell Scientific.

Datta, S. (1992). Effects of the availability of allies on female dominance structure. In *Coalitions and alliances in humans and other animals* (ed. A. Harcourt and F. de Waal), pp. 61–82, Oxford: Oxford Univ. Press.

Darwin, C. (1859). *On the origin of species.* London: J. Murray.

Darwin, C. (1871). *The descent of man and selection in relation to sex.* London: J. Murray.

Davies, N. B. (1976). Food, flocking and territorial behavior in the pied wagtail *Motacilla alba* in winter. *J. Anim. Ecol.*, **45**, 235–54.

Davies, N. B. (1992). *Dunnock behavior and social evolution.* Oxford: Oxford Univ. Press.

Davies, N. B., and Houston, A. I. (1981). Owners and satellites: the economics of territory defence in the pied wagtail *Motacilla alba*. *J. Anim. Ecol.*, **50**, 157–80.

Davis, L. S. (1984). Alarm calling in Richardson's ground squirrels (*Spermophilus richarsoni*). *Z. Tierpsychol.*, **66**, 152–64.

Dawkins, R. (1976). *The selfish gene.* 1st ed. Oxford: Oxford Univ. Press.

Dawkins, R. (1989). *The selfish gene.* 2nd ed. Oxford: Oxford Univ. Press.

Deag, J., and Crook, J. (1971). Social behavior and "agnostic buffering" in the wild barbary macaque. *Folia primatol*, **15**, 183–200.

DeNault, L. K., and McFarlane, D. A. (1995). Reciprocal altruism between male vampire bats, *Desmodus rotundus*. *Anim. Behav.*, **49**, 855–56.

Dennett, D. C. (1987). *The intentional stance*. Cambridge: MIT/Bradford Books.

Dittus, W. P. (1979). The evolution of behavior regulating density and age-specific distributions in a primate population. *Behaviour*, **69**, 265–302.

Dixson, A. F., Bossi, T., and Wickings, E. J. (1993). Male dominance and genetically determined reproductive success in the mandrill (*Mandrillus sphinx*). *Primates*, **34**, 525–32.

Dominey, W. J. (1983). Mobbing in colonially nesting fish, especially the bluegill, *Lepomis macrochirus*. *Copeia*, **1983**, 1086–88.

Donaldson, T. J. (1984). Mobbing behavior by *Stegastes albifasciatus* (Pomacentridae), a territorial mosaic damselfish. *Jap. J. Ichthy.*, **31**, 345–48.

Douglas-Hamilton, I., and Douglas-Hamilton, O. (1975). *Among the elephants*. Glasgow: Collins.

Drickamer, L. (1974). A ten-year summary of reproductive data for free-ranging *Macaca mulatta*. *Folia Primatol.*, **21**, 61–80.

Dublin, H. T. (1983). Cooperation and reproductive competition among female African elephants. In *Social Behavior of Female Vertebrates* (ed. S. Wasser), pp. 291–315, New York: Academic Press.

Dugatkin, L. A. (1988). Do guppies play Tit for Tat during predator inspection visits? *Behav. Ecol. and Sociobiol.*, **25**, 395–99.

Dugatkin, L. A. (1990). N-person games and the evolution of cooperation: a model based on predator inspection behavior in fish. *J. Theo. Biol.*, **142**, 123–35.

Dugatkin, L. A. (1991a). Predator inspection, Tit for Tat and shoaling: a comment on Masters and Waite. *Anim. Behav.*, **41**, 898–900.

Dugatkin, L. A. (1991b). Dynamics of the Tit for Tat strategy during predator inspection in guppies. *Behav. Ecol. Sociobiol.*, **29**, 127–32.

Dugatkin, L. A. (1992a). The evolution of the con artist. *Ethol. Sociobiol.*, **13**, 3–18.

Dugatkin, L. A. (1992b). Tendency to inspect predators predicts mortality risk in the guppy, *Poecilia reticulata*. *Behav. Ecol.*, **3**, 124–28.

Dugatkin, L. A. (1996). Tit for tat, byproduct mutualism and predator inspection: a reply to Connor. *Anim. Behav.* **51**, 455–57.

Dugatkin, L. A., and Alfieri. M., (1991a). Guppies and the Tit for Tat strategy: preference based on past interaction. *Behav. Ecol. Sociobiol.*, **28**, 243–46.

Dugatkin, L. A., and Alfieri, M., (1991b). Tit for Tat in guppies: the relative nature of cooperation and defection during predator inspection. *Evol. Ecol.*, **5**, 300–309.

Dugatkin, L. A., and Alfieri, M., (1992). Interpopulational differences in the use of the Tit for Tat strategy during predator inspection in the guppy. *Evol. Ecol.*, **6**, 519–26.

Dugatkin, L. A., Farrand, L., Wilkens, R., and Wilson, D. S. (1994). Altruism, Tit for Tat and "outlaw" genes. *Evol. Ecol.*, **8**, 431–37.

Dugatkin, L. A., and Godin, J.-G. J. (1992a). Prey approaching predators: a cost-benefit perspective. *Ann. Zool. Fennici*, **29**, 233–52.

Dugatkin, L. A., and Godin, J.-G. J. (1992b). Predator inspection, shoaling and foraging under predation hazard in the Trinidadian guppy. *Env. Biol. Fish.*, **34**, 265–75.

Dugatkin, L. A., Mesterton-Gibbons, M., and Houston, A. I. (1992). Beyond the Prisoner's Dilemma: towards models to discriminate among mechanisms of cooperation in nature. *Trends Ecol. Evol.*, **7**, 202–5.

Dugatkin, L. A., and Mesterton-Gibbons, M. (1995). Cooperation among unrelated individuals: reciprocal altruism, byproduct mutualism, and group selection in fishes. *Biosystems*, **37**, 19–30.

Dugatkin, L. A., and Reeve, H. K. (1993). Behavioral ecology and the "levels of selection": dissolving the group selection controversy. *Adv. Study Behav.*, **23**, 101–33.

Dugatkin, L. A., and Sih, A. (1995). Behavioral ecology and the study of partner choice. *Ethology*, **99**, 265–77.

Dugatkin, L. A., and Wilson, D. S. (1991). ROVER: A strategy for exploiting cooperators in a patchy environment. *Am. Nat.*, **138**, 687–701.

Dugatkin, L. A., and Wilson, D. S. (1992). The prerequisites of strategic behavior in the bluegill sunfish. *Anim. Behav.*, **44**, 223–30.

Dugatkin, L. A., and Wilson, D. S. (1993). Fish behaviour, partner choice and cognitive ethology. *Rev. Fish Biol. and Fisheries*, **3**, 368–72.

Dunford, C. (1977). Kin selection for ground squirrel alarm calls. *Am. Nat.*, **111**, 782–85.

Dunbar, R. (1980). Determinants and evolutionary consequences of dominance among female gelada baboons. *Behav. Ecol. Sociobiol.*, **7**, 253–65.

Dunbar, R. (1984a). Is social grooming altruistic? *Z. Tierpsychol.*, **64**, 163–72.

Dunbar, R. (1984b). Infant-use by male gelada in agnostic contexts: agonistic buffering, progeny protection or soliciting support? *Primates*, **25**, 28–35.

Dunbar, R. (1991). Functional significance of social grooming in primates. *Folia Primatol.*, **57**, 121–31.

Dunbar, R. I. M., and Dunbar, E. P. (1977). Dominance and reproductive success among female gelada baboons. *Nature*, **266**, 351–52.

East, M. (1981). Alarm calling and parental investment in the robin *Erihacus rubecula*. *Ibis*, **123**, 223–30.

Ekman, J. (1979). Coherence, composition and territories of winter social groups of the willow tit *Parus montanus* and the crested tit *P. cristatus*. *Ornis Scand.*, **10**, 56–68.

Ekman, J., and Askenmo, C. (1984). Social rank and habitat use in willow tit groups. *Anim. Behav.*, **32**, 508–14.

Ekman, J., and Askenmo, C. (1986). Reproductive cost, age-specific survival and a comparison of the reproductive strategy in two European tits (genus *Parus*). *Evol.*, **40**, 159–68.

Ekman, J., Cederholm, G., and Askenmo, C. (1981). Spacing and survival in winter groups of willow tits *Parus montanus* and crested tit *P. cristatus:* a removal experiment. *J. Anim. Ecol.*, **50**, 1–9.

Eldridge, N., and Gould, S. J. (1972). Puncuated equilibrium: an alternative to phyletic gradualism. In *Models of Paleobiology* (ed. T. J. Schopf), pp. 82–115, San Francisco: Freeman, Cooper.

Elgar, M. (1986). House sparrows establish foraging flocks by giving chirrup calls if the resource is divisible. *Anim. Behav.*, **34**, 169–74.

Ellis, D. H., Bednarz, J. C., Smith, D. G., and Flemming, S. P. (1993). Social foraging classes in raptorial birds. *BioScience*, **43**, 14–20.

Else, J., and Lee, P. (eds.) (1986). *Primate ontogeny, cognition and social behavior.* Cambridge: Cambridge Univ. Press.

Emerson, A. E. (1939). Social coordination and the superorganism. *Am. Mid. Nat.*, **21**, 182–209.

Emerson, A. E. (1946). The biological basis of social cooperation. *Ill. Acad. Sci. Trans.*, **39**, 9–18.

Emerson, A. E. (1960). The evolution of adaptation in population systems. In *Evolution after Darwin* (ed. S. Tax), pp. 307–48, Chicago: Chicago Univ. Press.

Emlen, S. T. (1984). Cooperative breeding in birds and mammals. In *Behavioural ecology*, 2nd ed. (ed. J.R. Krebs and N. B. Davies), pp. 305–39, Sunderland: Sinauer Assoc.

Emlen, S. T. (1991). Evolution of cooperative breeding in birds and mammals. In *Behavioural ecology*, 3rd ed. (ed. J. R. Krebs and N. B. Davies), pp. 301–37, Oxford: Blackwell Scientific.

Endler, J. (1986). *Natural selection in the wild*. Princeton, NJ: Princeton Univ. Press.

Enquist, M., and Leimar, O. (1993). The evolution of cooperation in mobile organisms. *Anim. Behav.*, **45**, 747–57.

Eshel, I. (1972). On the neighbor effect and the evolution of altruistic traits. *Theo. Pop. Biol.*, **3**, 258–77.

Eshel, I., and Cavalli-Sforza, L. L. (1982). Assortment of encounters and the evolution of cooperation. *Proc. Natl. Acad. Sci. USA*, **79**, 1331–35.

Eshel, I., and Weinshall, D. (1988). Cooperation in a repeated game with random payment functions. *J. Appl. Prob.*, **25**, 478–91.

Espinas, A. V. (1878). *Des societes animales*. Paris, Bailliere.

Essapian, F. S. (1963). Observations on abnormalities of parturition in captive bottle-nosed dolphins. *J. Mamm.*, **44**, 405–14.

Estes, R. D., and Goddard, J. (1967). Prey selection and hunting behavior of the African dog. *J. Wildlife Manag.*, **31**, 52–70.

Evans, H. E. (1977). Extrinsic versus intrinsic factors in the evolution of insect sociality. *Biosci.*, **27**, 613–17.

Evans, C. S., Evans, L., and Marler, P. (1993). On the meaning of alarm calls: functional reference in an avian vocal system. *Anim. Behav.*, **46**, 23–38.

Evans, C. S., and Marler, P. (1994). Food calling and audience effects in male chickens, *Gallus gallus:* their relationships to food availability, courtship and social facilitation. *Anim. Behav.*, **47**, 1159–70.

Evans, R. M. (1982). Foraging-flock recruitment at a black-billed gull colony: implications for the information center hypothesis. *Auk*, **99**, 24–30.

Evans, R. M., and Welham, C. V. J. (1985). Aggregative mechanisms and behavior in ring-billed gulls departing from a colony. *Can. J. Zool.*, **63**, 2767–74.

Ewer, R. F. (1973). *The carnivores*. Ithaca: Cornell Univ. Press.

Fairbanks, L. A. (1980). Relationships among adult females in captive vervet monkeys: testing a model of rank-related attractiveness. *Anim. Behav.*, **28**, 853–59.

Fairbanks, L. A. (1990). Reciprocal benefits of allomothering for female vervet monkeys. *Anim. Behav.*, **40**, 553–62.

Fairbanks, L. A., and McGuire, M. M. (1984). Determinants of fecundity and reproductive success in captive vervet monkeys. *Am. J. Primatol.*, **7**, 27–38.

Falls, J. B. (1982). Individual recognition by song in birds. In *Acoustic communication in birds* (ed. D. Kroodsma and E. Miller), pp. 237–78, New York: Academic Press.

Farrell, J., and Ware, R. (1989). Evolutionary stability in the repeated Prisoner's Dilemma. *Theo. Pop. Biol.*, **36**, 161–68.

Faulkes, C. G., Abbott, D. H., Jarvis, J., and Sherriff, F. E. (1989). Social suppression of reproduction in male naked mole rats, *Heterocephalus glaber*. *J. Reprod. Fertil. Abstr. Ser.*, **3**, 113.

Faulkes, C. G., Abbott, D. H., and Jarvis, J. (1990). Social suppression of ovarian cyclicity in captive and wild colonies of the naked mole rat, *Heterocephalus glaber*. *J. Reprod. Fertil.*, **88**, 559–68.

Faulkes, C. G., Abbott, D. H., Liddell, C. E., George, L. M., and Jarvis, J. (1991). Hormonal and behavioral aspects of reproductive suppression in female naked mole-rats. In *The biology of the naked mole-rat* (ed. P. W. Sherman, J. Jarvis, and R. D. Alexander), pp. 426–45, Princeton, NJ: Princeton Univ. Press.

Feldman, M., and Thomas, E. (1987). Behavior dependent contexts for the repeated plays of the Prisoner's Dilemma II: Dynamical aspects of the evolution of cooperation. *J. Theo. Biol.*, **128**, 297–315.

Ferriere, R., and Michod, R. E. (1995). Invading wave of cooperation in a spatially iterated prisoner's dilemma. *Proc. R. Soc. Lond.*, **259**, 77–83.

Ficken, M. S., Witkin, S. R., and Weise, C. M. (1981). Associations among members of flocks of a black-capped chickadee flock. *Behav. Ecol. Sociobiol.* **8**, 245–50.

Fischer, E. A. (1980). The relationship between mating system and simultaneous hermaphroditism in the coral reef fish, *Hypoplectrus nigricans* (Serranidae). *Anim. Behav.*, **28**, 620–33.

Fischer, E. A. (1984). Egg trading in the chalk bass, *Serranus tortugarum,* a simultaneous hermaphrodite. *Z. Tierpsychol.*, **66**, 143–51.

Fischer, E. A. (1986). Mating systems of simultaneously hermaphroditic Serranid fishes. In *Indo-Pacific fish biology: proceedings of the Second Conference on Indo-Pacific Fishes* (ed. T. Uyeno, R. Arai, T. Taniuchi, and K. Matsuura), pp. 776–84, Tokyo: Icthyological Society of Japan.

Fischer, E. A. (1988). Simultaneous hermaphroditism, Tit for Tat, and the evolutionary stability of social systems. *Ethol. Sociobiol.*, **9**, 119–36.

Fisher, J. (1954). Evolution and bird sociality. In *Evolution as a process* (ed. J. Huxley, A. C. Hardy, and E. B. Ford), pp. 71–83, London: Allen and Unwin.

Fisher, R. A. (1930). *The genetical theory of natural selection.* 1st ed. New York: Dover.

Fisher, R. A. (1958). *The genetical theory of natural selection.* 2nd ed. New York: Dover.

FitzGerald, G., and van Havre, N. (1987). The adaptive significance of cannibalism in sticklebacks. *Behav. Ecol. Sociobiol.,* **20**, 125–28.

FitzGibbon, C. D., and Fanshawe, J. H. (1988). Stotting in Thomson's gazelles: an honest signal of condition. *Behav. Ecol. Sociobiol.*, **23**, 69–74.

FitzGibbon, C. D. (1994). The costs and benefits of predator inspection behaviour in Thomson's gazelles. *Behav. Ecol. Sociobiol.*, **34**, 139–48.

Fletcher, D. J., and Ross, K. G. (1985). Regulation of reproduction in eusocial hymenoptera. *Ann. Rev. Entomol.*, **30**, 319–43.

Foster, S. A. (1985a). Group foraging by a coral reef fish: a mechanism for gaining access to defended resources. *Anim. Behav.,* **33**, 782–92.

Foster, S. A. (1985b). Size-dependent territory defense by a damselfish: a determinant of resource use by group-foraging surgeonfishes. *Oecologia*, **647**, 499–505.

Foster, S. A. (1987). Acquisition of a defended resource: a benefit of group foraging for the neotropical wrasse, *Thalassoma lucasanum. Env. Biol. Fishes*, **19**, 215–22.

Foster, S. A., and Ploch, S. (1990). Determinants of variation in antipredator behavior of territorial male threespine sticklebacks in the wild. *Ethology*, **84**, 281–94.

Frankenberg, E. (1981). The adaptive significance of avian mobbing: IV. "alerting others" and "perception advertisement" in blackbirds facing an owl. *Z. Tierpsychol.*, **55**, 97–118.

Frean, M.R. (1994). The prisoner's dilemma without synchrony. *Proc. R. Soc. Lond, Series B.*, **257**, 75–79.

Freeland, W. J. (1976). Pathogens and the evolution of primate sociality. *Biotropica*, **8**, 12–24.

French, D. P. (1980). Cleaning symbiosis in sunfish hybrids under laboratory conditions. *Copeia*, **1980**, 869–70.

Friedman, J. W., and Hammerstein, P. (1991). To trade, or not to trade, that is the question. In *Game equilibrium models* (ed. R. Selten), pp. 257–75, New York: Springer-Verlag.

Frumhoff, P. C., and Baker, J. (1988). A genetic component to division of labor within honeybee colonies. *Nature*, **333**, 358–61.

Gadagkar, R. (1985a). Evolution of insect sociality—a review of some attempts to test modern theories. *Proc. Indian Acad. Sci.*, **94**, 309–24.

Gadagkar, R. (1985b). Kin recognition in social insects and other animals—a review of recent findings and a consideration of their relevance for the theory of kin selection. *Proc. Indian Acad. Sci.*, **94**, 587–621.

Gadagkar, R. (1990a). Evolution of eusociality: the advantages of assured fitness returns. *Phil. Trans. R. Soc. Lond.*, **329**, 17–25.

Gadagkar, R. (1990b). The haplodiploidy threshold and social evolution. *Current Science*, **59**, 374–76.

Gadagkar, R. (1990c). Origin and evolution of eusociality: a perspective from studying primitively eusocial wasps. *Genetics*, **69**, 113–25.

Gadagkar, R. (1991a). Demographic predisposition to the evolution of sociality: a hierarchy of models. *Proc. Natl. Acad. Sci. USA*, **88**, 10993–97.

Gadagkar, R. (1991b). On testing the role of genetic asymmetries created by haplodiploidy in the evolution of eusociality in the Hymenoptera. *J. Genetics*, **70**, 1–31.

Gadagkar, R. (1994). Why the definition of eusociality is not helpful to its evolution and what we should do about it. *Oikos*, **70**, 485–88.

Gadgil, M. (1975). Evolution of social behavior through interpopulational selection. *Proc. Natl. Acad. Sci. USA*, **72**, 1199–1201.

Gaffrey, G. (1957). Zur fortpflanzungbsbiologie bei hunden. *Zool. Gart. Jena*, **23**, 251.

Gamboa, G. J. (1988). Sister, aunt-niece, and cousin recognition by social wasps. *Behav. Genet.*, **18**, 409–23.

Gamboa, G. J., Reeve, H. K., and Pfennig, D. W. (1986). The evolution and ontogeny of nestmate recognition in social wasps. *Ann. Rev. Entomol.*, **31**, 431–54.

George, C. (1960). Behavioral interactions in the pickerel and the mosquitofish. Ph.D. dissertation. Harvard University.

Getty, T. (1987). Dear enemy and the Prisoner's Dilemma: why should territorial neighbors form defensive coalitions. *Amer. Zool.*, **27**, 327–36.

Gittleman, J. L. (1985). Functions of communal care in mammals. In *Evolution—essays in honour of John Maynard Smith* (ed. P. J. Greenwood, P. H. Harvey, and M. Slatkin), pp. 187–205. London: Cambridge Univ. Press.

Gittleman, J. L., and Thompson, S. D. (1988). Energy allocation in mammalian reproduction. *Am. Zool.*, **28**, 863–75.

Godard, R. (1991). Long-term memory of neighbors in a migratory songbird. *Nature*, **350**, 228–29.

Godard, R. (1993). Tit-for-tat among neighboring hooded warblers. *Behav. Ecol. Sociobiol.*, **33**, 45–50.

Godfray, H. C. J., and Grafen, A. (1988). Unmatedness and the evolution of eusociality. *Am. Nat.*, **131**, 303–5.

Godin, J.-G. (Ed.). (1996). *Behavioral ecology of fishes*. Oxford: Oxford Univ. Press.

Godin, J.-G., and Davis, S. A. (1995a). Who dares, benefits: predator inspection behaviour in the guppy (*Poecilia reticulata*) deters predator pursuit. *Proc. R. Soc. Lond., Series B.*, **259**, 193–200.

Godin, J.-G., and Davis, S. A. (1995b). Boldness and predator deterrence: a reply to Milinski and Bolthauser. *Proc. R. Soc. Lond. B.*, **262**, 107–12.

Godin, J. G., and Dugatkin, L. A. (1996). Female mating preference for bold males in the guppy. *Proc. Natl. Acad. Sci. USA*, **93**, 10262–10267.

Godin, J., and Smith, S. (1989). A fitness cost of foraging in the guppy. *Nature*, **333**, 69–71.

Goodall, J. (1986a). Social rejection, exclusion and shunning among Gombe chimpanzees. *Ethol. Sociobiol.*, **7**, 227–36.

Goodall, J. (1986b). *The chimpanzee of Gombe: patterns of behavior.* Cambridge, MA: Belknap Press.

Gori, D. F. (1988). Colony-facilitated foraging in yellow-headed blackbirds: experimental evidence for information transfer. *Ornis Scand.*, **19**, 224–30.

Gorlick, D.L., Atkins, P. D., and Losey, G. S. (1978). Cleaning stations as water holes, garbage dumps and sites for the evolution of reciprocal altruism. *Am. Nat.*, **112**, 341–53.

Gotmark, F. (1990). A test of the information-centre hypothesis in a colony of sandwich terns *Sterna sandvicensis. Anim. Behav.*, **39**, 487–95.

Gouzoules, H., Gouzoules, S., and Fedigan, L. (1982). Behavioral dominance and reproductive success in female Japanese macaques. *Anim. Behav.*, **30**, 1138–50.

Gouzoules, S. (1984). Primate mating systems, kin associations and cooperative behavior: evidence for kin recognition. *Yrbk. Phys. Anthropol.*, **27**, 99–134.

Gouzoules, S., and Gouzoules, H. (1987). Kinship. In *Primate societies* (ed. B. Smuts, D. L. Cheney, R. Seyfarth, R. Wrangham, and T. Struhsaker), pp. 299–305, Chicago: Univ. of Chicago Press.

Grafen, A. (1979). The hawk-dove game played between relatives. *Anim. Behav.*, **27**, 905–7.

Grafen, A. (1984). Natural selection, kin selection and group selection. In *Behavioural ecology*, 2nd ed. (ed. J. Krebs and N. Davies), pp. 62–84. London: Sinauer Assoc.

Grafen, A. (1985). A geometric view of relatedness. *Oxford Surveys in Evol. Biol.*, **2**, 28–89.

Greenberg, L. (1988). Kin recognition in the sweat bee, *Lasioglossum zephyrum. Behav. Genet.*, **18**, 425–37.

Griffin, D. R. (1984). *Animal thinking.* Cambridge, MA: Harvard Univ. Press.

Grim, P. (1995). The greater generosity of the spatialized prisoner's dilemma. *J. Theo. Biol.*, **173**, 353–59.

Grim, P. (1996). Spatalization and greater generosity in the stochastic Prisoner's Dilemma. *Biosystems*, in press.

Grinnell, J., Packer, C., and Pusey, A. (1995). Cooperation in male lions: kinship, reciprocity or mutualism? *Anim. Behav.*, **49**, 95–105.

Gross, M. (1994). The evolution of behavioral ecology. *Trends Ecol. Evol.*, **9**, 358–60.

Guhl, A., Collias, N., and Allee, W. C. (1945). Mating behavior and the social hierarchy in small flocks of white leghorns. *Physiol. Zool.*, **18**, 365–90.

Guthrie, R. D. (1971). A new theory of mammalian rump patch evolution. *Behaviour*, **38**, 132–45.

Gyger, M., Karakashian, S. J., and Marler, P. (1986). Avian alarm calling: is there an audience effect? *Anim. Behav.*, **34**, 1570–72.

Hagen, R., Smith, R., and Rissing, S. (1988). Genetic relatedness among cofoundresses in two desert ant species. *Psyche*, **95**, 191–201.

Haldane, J. B. S. (1932). *The causes of evolution.* London: Longmans Green.

Haldane, J. B. S. (1955). Population genetics. *New Biology,* **18,** 34–51.

Hall, K. R. (1960). Social vigilance behavior of the chacma baboon, *Papio anubis. Behaviour,* **16,** 261–94.

Hamilton, W. D. (1963). The evolution of altruistic behavior. *Am. Nat.,* **97,** 354–56.

Hamilton, W. D. (1964a). The genetical evolution of social behaviour, I. *J. Theo. Biol.,* **7,** 1–16.

Hamilton, W. D. (1964b). The genetical evolution of social behaviour. II. *J. Theo. Biol.,* **7,** 17–52.

Hamilton, W. D. (1967). Extraordinary sex ratios. *Science,* **156,** 477–87.

Hamilton, W. D. (1971). Geometry of the selfish herd. *J. Theo. Biol.,* **31,** 295–311.

Hamilton, W. D. (1972). Altruism and related phenomena, mainly in social insects. *Ann. Rev. Ecol. System.,* **3,** 192–232.

Hamilton, W. D. (1975). Innate social attitudes in man: an approach from evolutionary genetics. In *Biosocial anthropology* (ed. R. Fox), pp. 133–55, New York: Wiley.

Hand, J. (1986). Resolution of social conflicts: dominance, egalitarianism, spheres of influence and game theory. *Q. Rev. Biol.,* **61,** 201–20.

Hansell, M. (1987). Nest building as a facilitating and limiting factor in the evolution of eusociality in Hymenoptera. *Oxford Surveys Evol. Biol.,* **4,** 155–81.

Harcourt, A. H., and Stewart, K. J. (1987). The influence of help in contests on dominance rank in primates; hints from gorillas. *Anim. Behav.,* **35,** 182–90.

Harcourt, A. H., and Stewart, K. J. (1989). Functions of alliances in contests within wild gorilla groups. *Behaviour,* **109,** 176–90.

Harcourt, A. H., and de Waal, F. B. M. (eds.). (1992). *Coalitions and alliances in humans and other animals.* Oxford: Oxford Univ. Press.

Harpending, H., and Sobus, J. (1987). Sociopathy as an adaptation. *Ethol. Sociobiol.,* **8,** 63s–72s.

Hart, B. L., and Hart, L. (1992). Reciprocal allogrooming in impala. *Anim. Behav.,* **44,** 1073–83.

Hart, B. L., Hart, L., Mooring, M. S., and Olubayo, R. (1992). Biological basis of grooming behavior in the antelope: the body-size, vigilance and habitat principles. *Anim. Behav.,* **44,** 615–31.

Harre, R., and Reynolds, R. (eds.) (1984). *The meaning of primate signals.* Cambridge: Cambridge Univ. Press.

Harvey, P. H., and Greenwood, P. J. (1978). Anti-predator defence strategies: some evolutionary problems. In *Behavioral ecology,* 1st ed. (ed. J. Krebs and N. Davies), pp. 129–51, Sunderland, MA: Sinauer.

Harvey, P. H., and Pagel, M. D. (1991). *The comparative method in evolutionary biology.* Oxford: Oxford Univ. Press.

Hauser, M. (1988). How infant vervet monkeys learn to recognize starling alarm calls: the role of experience. *Behaviour,* **105,** 187–201.

Hauser, M., Cheney, D., and Seyfarth, R. (1986). Group extinction and fusion in free-ranging vervet monkeys. *Am. J. Primat.,* **11,** 63–77.

Hausfater, G. (1975). *Dominance and reproduction in baboons.* Basel: Krager.

Healy, S. (1992). Optimal memory: toward an evolutionary ecology of animal cognition. *Trends Ecol. Evol.,* **8,** 399–400.

Hector, D. (1986). Cooperative hunting and its relationship to foraging success and prey size in an avian predator. *Ethology,* **73,** 247–57.

Heinrich, B. (1981). The mechanisms and energetics of honeybee swarm regulation. *J. Exp. Biol.*, **85**, 61–87.

Heinrich, B. (1985). Temperature regulation in honeybees. In *Experimental behavioral ecology and sociobiology* (ed. B. Holldobler and M. Lindauer), pp. 393–406, Sunderland, MA: Sinauer.

Heinrich, B. (1987). Thermoregulation by individual honeybees. In *Neurobiology and behavior in honeybees* (ed. R. Menzel and A. Mercer), pp. 102–111, Berlin: Springer-Verlag.

Heinrich, B. (1988a). Food sharing in the raven, *Corvus corax*. In *The ecology of social behavior* (ed. C. N. Slobodchikoff), pp. 285–311, San Diego, CA: Academic Press.

Heinrich, B. (1988b). Winter foraging at carcasses by three sympatric corvids, with emphasis on recruitment by the raven, *Corvus corax*. *Behav. Ecol. Sociobiol.*, **23**, 141–56.

Heinrich, B. (1989). *Ravens in winter.* New York: Simon and Schuster.

Heinrich, B., and Marzluff, J. M. (1991). Do common ravens yell because they want to attract others? *Behav. Ecol. Sociobiol.*, **28**, 13–21.

Heinsohn, R., and Packer, C. (1995). Complex cooperative strategies in group-territorial African lions. *Science*, **269**, 1260–1263.

Helfman, G. (1989). Threat-sensitive predator avoidance in damselfish-trumpetfish interactions. *Behav. Ecol. Sociobiol.*, **24**, 47–58.

Hellmich, R. L., Kulincevic, J. M., and Rothenbluhler, W. C. (1985). Selection for low and high pollen hoarding honeybees. *J. Hered.*, **76**, 155–58.

Hemelrijk, C. K. (1990a). A matrix partial correlation test used in investigations of reciprocity and other social interaction patterns at the group level. *J. Theo. Biol.*, **143**, 405–20.

Hemelrijk, C. K. (1990b). Models of, and tests for, reciprocity, undirectionality and other social interaction patterns at a group level. *Anim. Behav.*, **39**, 1013–29.

Hemelrijk, C. K. (1991). Interchange of "altruistic" acts as an epiphenomenon. *J. Theo. Biol.*, **153**, 137–39.

Hemelrijk, C. K., and Ek, A. (1991). Reciprocity and interchange of grooming and "support" in captive chimpanzees. *Anim. Behav.*, **41**, 923–35.

Hennessy, D. F. (1986). On the deadly risk of predator harassment. *Ethology*, **72**, 72–74.

Hepper, P. G. (Ed.). (1991). *Kin recognition.* Cambridge: Cambridge Univ. Press.

Hert, E. (1985). Individual recognition of helpers by breeders in the cichlid fish *Lamproglus brichardi*. *Z. Tierpsychol.*, **68**, 313–25.

Hews, D. K. (1988). Alarm response in larval western toads, *Bufo boreas:* a release of larval chemical by a natural predator and its effect on predator capture efficiency. *Anim. Behav.*, **36**, 125–33.

Hines, W., and Maynard Smith, J. (1979). Games between relatives. *J. Theo. Biol.*, **79**, 19–30.

Hirshleifer, D., and Rasmussen, E. (1989). Cooperation in a repeated prisoner's dilemma game with ostracism. *J. Econ. Behav. and Organization*, **12**, 87–106.

Hirshleifer, J., and Martinez-Coll, J. C. (1988). What strategies can support the evolutionary emergence of cooperation? *J. Conflict Resol.*, **32**, 367–98.

Hirth, D. H., and McCullough, D. (1977). Evolution of alarm signals in ungulates with special reference to white-tailed deer. *Am. Nat.*, **111**, 31–42.

Hobbes, T. (1651). *Leviathan.* Reprinted by MacMillan Publishers (1962).

Hogstad, O. (1987). Social rank in winter flocks of willow tits. *Ibis*, **129**, 1–9.

Hogstad, O. (1995). Alarm calling by willow tits, *Paras montanus,* as mate investment. *Anim. Behav.*, **49**, 221–25.

Holldobler, B., and Wilson, E. O. (1977). The number of queens: an important trait in ant evolution. *Naturwissenschaften*, **64**, 8–15.

Holldobler, B., and Wilson, E. O. (1990). *The ants.* Cambridge, MA: Harvard Univ. Press.

Hoogland, J. L. (1983). Nepotism and alarm calls in the black-tailed prairie dog (*Cynomys ludovicianus*). *Anim. Behav.*, **31**, 472–79.

Hoogland, J. L. (1995). *The black-tailed prairie dog.* Chicago: Univ. of Chicago Press.

Horrocks, J., and Hunte, W. (1983). Maternal rank and offspring rank in vervet monkeys: an appraisal of the mechanisms of rank acquisition. *Anim. Behav.*, **31**, 772–82.

Houston, A., Schmid-Hempel, P., and Kacelnik, A. (1988). Foraging strategy, worker mortality, and the growth of the colony in social insects. *Am. Nat.*, **131**, 107–14.

Howe, N. R., and Sheik, Y. M. (1975). Anthopleurine: a sea anemone alarm pheromone. *Science,* **189**, 386–88.

Hrdy, S. B. (1976). Care and exploitation of non-human primate infants by conspecifics other than the mother. *Adv. Study of Behav.*, **6**, 101–58.

Hughes, C., Queller, D. C., Strassman, J., and Davis, S. (1993). Relatedness and altruism in Polistes wasps. *Behav. Ecol.*, **4**, 128–37.

Huntingford, F. A. (1984). Some ethical issues raised by studies of predation and aggression. *Anim. Behav.*, **32**, 210–15.

Hutchins, D., and Barash, D. (1976). Grooming in primates: implications for its utilitarian function. *Primates*, **17**, 145–50.

Hutson, V. C. L., and Vickers, G. (1995). The spatial struggle of tit-for-tat and defect. *Phil. Trans. R. Soc. Lond. B*, **348**, 393–404.

Huxley, T. H. (1888). The struggle for existence: a programme. *Nineteenth Century,* **23**, 161–81.

Isbell, L. A. (1991). Contest and scramble competition: patterns of female aggression and ranging behavior among primates. *Behav. Ecol.*, **2**, 143–55.

Ishihara, M. (1987). Effect of mobbing toward predators by the damselfish, *Pomacentrus coelestis* (Pisces: Pomacentridae). *J. Ethol.,* **5**, 43–52.

Janzen, D. H. (1970). Altruism by coatis in the face of predation by boa constrictor. *J. Mamm.*, **51**, 387–89.

Jarman, M. V. (1979). *Impala social behavior: territory, hierarchy, mating and the use of space.* Berlin: Springer-Verlag.

Jarvis, J. (1981). Eusociality in a mammal: cooperative breeding in the naked mole-rat. *Science*, **212**, 571–73.

Jarvis, J., O'Riain, M. J., Bennett, N. C., and Sherman, P. W. (1994). Mammalian sociality: a family affair. *Trends Ecol. Evol.*, **9**, 47–51.

Jeanne, R. L. (1972). Social biology of the neotropical wasp *Mischocyttarus drewseni*. *Bull. Mus. Comp. Zool., Harvard Univ.*, **144**, 63–150.

Jeanne, R. L. (1980). Evolution of social behavior in the vespidae. *Ann. Rev. Entomol.*, **25**, 371–96.

Jennions, M. D., and MacDonald, D. W. (1994). Cooperative breeding in mammals. *Trends Ecol. Evol.*, **9**, 89–93.

Joshi, N. V. (1987). Evolution of reciprocation in structured demes. *J. of Genetics*, **1**, 69–84.

Joshi, N. V., and Gadagkar, R. (1985). Evolution of sex ratios in hymenoptera: kin selection, local mate competition, polyandry and kin recognition. *J. of Genetics,* **64,** 41–58.

Judge, P. G. (1994). Intergroup relations between alpha females in a population of free-ranging rhesus macaques. *Folia Primatol.,* **63,** 63–70.

Kauffman, J. H. (1965). A three year study of mating behavior in a free- ranging band of rhesus monkeys. *Ecology,* **46,** 500–512.

Keller, L. (1991). Queen number, mode of colony founding and queen reproductive success in ants. *Ethol. Ecol. Evol.,* **3,** 307–16.

Keller, L. (ed.) (1993). *Queen number and sociality in insects.* Oxford: Oxford Univ. Press.

Keller, L. (1995a). Social life: the paradox of multiple-queen colonies. *Trends Ecol. Evol.,* **10,** 355–60.

Keller, L. (1995b). Quantifying the level of eusociality. *Proc. R. Soc. Lond., Series B,* **260,** 311–15.

Keller, L., and Reeve, H. K. (1994). Partitioning of reproduction in animal societies. *Trends Ecol. Evol.,* **9,** 98–102.

Keverne, E. B., Martenz, N. D., and Tuite, B. (1989). Beta-endorphin concentrations in cerebrospinal fluid of monkeys are influenced by grooming relationships. *Psychoneuroendocrinology,* **14,** 155–61.

Kiis, A., and Moller, A. P. (1986). A field test of information transfer in communally nesting greenfinches *Carduelis chloris. Anim. Behav.,* **34,** 1251–55.

Kingdon, J. (1979). *East African mammals: an atlas of evolution in Africa.* New York: Academic Press.

Kitchen, W. D. (1972). The social behavior and ecology of the pronghorn. Ph.D. dissertation. Univ. of Michigan.

Klahn, J. E. (1979). Philopatric and nonphilopatric foundress associations in the social wasp *Polistes fuscatus. Behav. Ecol. Sociobiol.,* **5,** 417–24.

Klump, G., and Shalter, M. (1984). Acoustic behaviour of birds and mammals in the predator context. *Z. Tierpsychol.,* **66,** 189–226.

Knight, S. K., and Knight, R. L. (1983). Aspects of food finding by wintering bald eagles. *Auk,* **100,** 477–84.

Koenig, W. D. (1988). Reciprocal altruism in birds: a critical review. *Ethol. Sociobiol.,* **9,** 73–84.

Koenig, W. D., and Mumme, R. L. (1987). *Population ecology of the cooperatively breeding acorn woodpecker.* Princeton: Princeton Univ. Press.

Koivula, K., and Orwekk, M. (1988). Social rank and winter survival in the willow tit *Parus montanus. Ornis Fenn.,* **65,** 114–20.

Konishi, M. (1963). The role of auditory feed-back in the vocal behaviour of domestic fowl. *Z. Tierpsychol.,* **20,** 349–67.

Krebs, J. (1982). Territorial defence in the great tit *(Parus major):* do residents always win? *Behav. Ecol. Sociobiol.,* **11,** 185–94.

Krebs, J. R. (1974). Colonial nesting and social feeding as strategies for exploiting food resources in the great blue heron *(Ardea herodias). Behaviour,* **51,** 99–134.

Kropotkin, P. (1908). *Mutual Aid.* 3rd ed. London: William Heinemann.

Kruijt, J. P. (1964). Ontogeny of social behaviour in Burmese red jungle fowl *(Gallus gallus spadiceus)* Bonaterre. *Behav. Supplement,* **12,** 1–201.

Kukuk, P. F., Eickwort, G. C., Ravaret-Ricther, M., Alexander, B., Gibson, R., Morse, R., and Ratnieks, F. L. W. (1989). The importance of the sting in the evolution of sociality in the hymenoptera. *Ann. Ent. Soc. Am.,* **82,** 1–5.

Kummer, H. (1978). On the value of social relationships to nonhuman primates: a heuristic scheme. *Soc. Sci. Inform.*, **12**, 687–705.

Kunz, T. H., and Allgaier, A. L. (1994). Allomaternal care: helper-assisted birth in the Rodrigues fruit bat, *Pteropus rodricensis* (Chiroptera: Pteropodidae). *J. Zool., Lond.*, **232**, 691–700.

Kurland, J. A. (1977). *Kin selection in the Japanese monkey.* Basel: Karger.

Kynard, B. (1979). Breeding behaviour of a lacustrine population of threespine sticklebacks. *Behaviour*, **67**, 178–207.

Lacey, E. A., Alexander, R. D., Braude, S. H., Sherman, P. W., and Jarvis, J. (1991). An ethogram for the naked mole-rat: non-vocal behaviors. In *The biology of the naked mole-rat* (ed. P. W. Sherman, J. Jarvis, and R. D. Alexander), pp. 209–42, Princeton, NJ: Princeton Univ. Press.

Lacey, E. A., and Sherman, P. W. (1991). Social organization of naked mole-rats: evidence for divisions of labor. In *The Biology of The Naked Mole-Rat,* (ed. P. W. Sherman, J. Jarvis, and R. D. Alexander), pp. 275–336, Princeton, NJ: Princeton Univ. Press.

LaGory, K. E. (1987). The influence of habitat and group characteristics on the alarm and flight response of white-tailed deer. *Anim. Behav.*, **35**, 20–25.

Lawick-Goodall, J. (1968). The behaviour of free-living chimpanzees in the Gombe Stream Reserve. *Anim. Behav. Monographs*, **1**, 161–311.

Lazarus, J. and N. Metcalfe. (1990). Tit for Tat cooperation in sticklebacks: a critique of Milinski. *Anim. Behav.*, **39**, 987–89.

Lee, P. C. (1983). Context-specific unpredictability in dominance relationships. In *Primate social relationships,* (ed. R. Hinde), pp. 35–44, Oxford: Blackwell Scientific.

Lee, P. C. (1987). Allomothering among African elephants. *Anim. Behav.*, **35**, 278–91.

Lefebrve, B. (1982). Food exchange strategies in an infant chimpanzee. *J. Human Evol.*, **11**, 195–204.

Leffelaar, D., and Robertson, R. (1984). Do male tree swallows guard their mates? *Behav. Ecol. Sociobiol.*, **16**, 73–79.

Leger, D. W., and Owings, D. H. (1978). Responses to alarm calls by California ground squirrels: effects of call structure and maternal status. *Behav. Ecol. Sociobiol.*, **3**, 177–86.

Leonard, J. L. (1990). The hermaphrodite's dilemma. *J. Theo. Biol.*, **147**, 361–72.

Leonard, J. L. (1993). Sexual conflict in simulataneous hermaphrodites: evidence from serranid fishes. *Env. Biol. Fish.*, **36**, 135–48.

Leopold, A. S. (1977). *The California quail.* Los Angeles: Univ. of California Press.

Leuthold, W. (1977). *African ungulates: a comparative review of their ethology and behavioral ecology.* Berlin: Springer-Verlag.

Levitt, P. R. (1975). General kin selection models for genetic evolution of sib altruism in diploid and haplodiploid species. *Proc. Natl. Acad. Sci. USA*, **72**, 4531–35.

Levin, B. R., and Kilmer, W. L. (1974). Interdemic selection and the evolution of altruism: a computer simulation. *Evolution*, **28**, 527–45.

Licht, T. (1989). Discriminating between hungry and satiated predators: the response of guppies *(Poecilia reticulata)* from high and low predation sites. *Ethology*, **82**, 238–43.

Ligon, J. D. (1991). Co-operation and reciprocity in birds and mammals. In *Kin recognition,* (ed. P. G. Hepper), pp. 30–59, Cambridge: Cambridge Univ. Press.

Lima, S. (1989). Iterated Prisoner's Dilemma: an approach to evolutionarily stable cooperation. *Am. Nat.*, **134**, 828–34.

Limbaugh, C. (1961). Cleaning symbiosis. *Sci. Amer.*, **205**, 167–78.

Lin, N., and Michener, C. D. (1972). Evolution of sociality in insects. *Q. Rev. Biol.*, **47**, 131–59.

Lindauer, M. (1948). Uber die Einwirkung von Duft- und Geschmacksstoffen sowie anderer Faktoren auf die Tanze der Bienen. *Z. Vergl. Physiol.*, **31**, 348–412.

Lindauer, M. (1952). Ein Beitrag zur Frage der Arbeitsteilung um Bienenstaat. *Z. Vergl. Physiol.*, **34**, 299–345.

Lindauer, M. (1954). Temperaturregulierung und Wasserhaushalt in Bienstaat. *Z. Vergl. Physiol.*, **36**, 391–432.

Litte, M. (1977). Behavioral ecology of the social wasps, *Mischocyttarus mexicanus*. *Behav. Ecol. Sociobiol.*, **2**, 229–46.

Loman, J., and Tamm, S. (1980). Do roosts serve as "information centres" for crows and ravens? *Am. Nat.*, **115**, 285–89.

Lombardo, M. (1985). Mutual restraint in tree swallows: A test of the Tit for Tat model of reciprocity. *Science*, **277**, 1363–65.

Lombardo, M. P. (1990). Tree swallows and Tit for Tat. *Ethol. Sociobiol.*, **11**, 521–28.

Lorberbaum, J. (1994). No strategy is evolutionarily stable in the repeated Prisoner's Dilemma. *J. Theo. Biol.*, **168**, 117–30.

Losey, G. S. (1972). The ecological importance of cleaning symbiosis. *Copeia*, **1972**, 820–23.

Losey, G. S. (1979). Fish cleaning symbiosis: proximate causes of host behavior. *Anim. Behav.*, **27**, 669–85.

Losey, G. S. (1987). Cleaning symbiosis. *Symbiosis*, **4**, 229–58.

Lucas, J. R., Creel, S. R., and Wasser, P. M. (1996). How to measure inclusive fitness, revisted. *Anim. behav.*, **51**, 225–228.

Lucas, N. S., Hume, E. M., and Henderson, H. (1937). On the breeding of the common marmoset *(Hapale jacchus)* in captivity when irradiated with ultra-violet rays. II. A ten-year family history. *Proc. Zool. Soc. Lond., Series A*, **107**, 205–11.

Lumsden, C. J. (1982). The social regulation of physical caste: the superorganism revisted. *J. Theo. Biol.*, **95**, 749–81.

McCourt, R. M., and Thomson, D. A. (1984). Cleaning behavior of the juvenile Panamic sergeant major, *Abudefduf troschelii* (Gill), with a resume of cleaning association in the gulf of California and adjacent waters. *Calif. Fish and Game*, **70**, 234–39.

McCracken, G. F., and Bradbury, J. W. (1981). Social organization and kinship in the polygynous bat *Phyllostomus hasatus*. *Behav. Ecol. Sociobiol.*, **8**, 11–34.

McCullough, D. R. (1969). The tule elk: its history, behavior and ecology. *Univ. Calif. Pub. Zool.*, **88**, 1–209.

MacDonald, D. W., and Moehlman, P. D. (1982). Cooperation, altruism and restraint in the reproduction of carnivores. In *Perspectives in Ethology,* (ed. P. Bateson and P. Klopfer), vol. 5, pp. 433–67, New York: Plenum Press.

Macedonia, J. M. (1990). What is communicated in the antipredator calls of lemurs: evidence from playback experiments with ringtailed and ruffed lemurs. *Ethology*, **86**, 177–90.

Macedonia, J. M., and Evans, C. S. (1993). Variation among mammalian alarm call signalling and the problem of meaning in animal signals. *Ethology*, **93**, 177–97.

Mackay-Sim, A., and Laing, D. G. (1982). Rats' responses to blood and body parts of stressed and non-stressed conspecifics. *Physiol. Behav.*, **27**, 503–10.

McKaye, K., and McKaye, N. (1977). Communal care and kidnapping of young by parental cichlids. *Evol.*, **31**, 674–81.

McKenna, J. (1979). The evolution of allomothering behavior among Colobine monkeys: function and opportunism in evolution. *Am. Anthropol.*, **81**, 818–40.

McKenna, J. (1981). Primate infant caregiving behavior: origins, consequences, and variability with emphasis on the common Indian langur monkey. In *Parental care in mammals,* (ed. D. J. Gubernick and P. H. Kloper), pp. 389–416, New York: Plenum Press.

McLean, I. G., and Rhodes, G. (1991). Enemy recognition and response in birds. In *Current ornithology,* (ed. D. Power), pp. 173–211, New York: Plenum Press.

McNab, B. K. (1973). Energetics and the distribution of vampires. *J. Mammology,* **54**, 131–144.

Macnair, M. R. (1978). An ESS for the sex ratio in animals with particular reference to the social hymenoptera. *J. Theo. Biol.,* **70**, 449–59.

Maestripieri, D. (1993). Vigilance costs of allogrooming in macaque mothers. *Am. Nat.,* **141**, 744–53.

Magurran, A. E. (1990). The adaptive significance of schooling as an antipredator defence in fish. *Ann. Zool. Fennici,* **27**, 51–66.

Magurran, A. E., and Girling, S. (1986). Predator recognition and response habituation in shoaling minnows. *Anim. Behav.,* **34**, 510–18.

Magurran, A. E., and Higham, A. (1988). Information transfer across fish shoals under predator threat. *Ethology,* **78**, 153–58.

Magurran, A. E., and Nowak, M. A. (1991). Another battle of the sexes: the consequences of sexual asymmetry in mating costs and predation risk in the guppy, *Poecilia reticulata. Proc. R. Soc. Lond. B,* **246**, 31–38.

Magurran, A. E. and Pitcher, T. J. (1987). Provenance, shoal size and the sociobiology of predator- evasion in minnow shoals. *Proc. R. Soc. Lond. B,* **229**, 439–65.

Major, P. F. (1978). Predator-prey interactions in two schooling species of fishes, *Caranx ignoblis* and *Stolephorus purpureus. Anim. Behav.,* **26**, 760–67.

Malcolm, J. R., and Marten, K. (1982). Natural selection and the communal rearing of pups in African wild dogs (*Lycaon pictus*). *Behav. Ecol. Sociobiol.,* **10**, 1–13.

Marler, P. (1955). Characteristics of some animal calls. *Nature,* **176**, 6–8.

Marler, P., Dufty, A., and Pickert, R. (1986a). Vocal communication in the domestic chicken: I. Does a sender communicate information about the quality of food referent to a receiver? *Anim. Behav.,* **34**, 188–93.

Marler, P., Dufty, A., and Pickert, R. (1986b). Vocal communication in the domestic chicken: II. Is a sender sensitive to the presence and absence of a receiver? *Anim. Behav.,* **34**, 194–98.

Marshall, J. C. (1970). The biology of communication in man and animals. In *New horizons in linguistics* (ed. J. Lyons), pp. 229–241, Harmondsworth: Penguin.

Marzluff, J. M., and Balda, R. P. (1992). *The pinyon jay: behavioral ecology of a colonial and cooperative corvid.* London: T. and A. D. Poyser.

Masters, M., and Waite, M. (1990). Tit-for tat during predator inspection or shoaling? *Anim. Behav.,* **39**, 603–4.

Matessi, C., and Jayakar, S. D. (1976). Conditions for the evolution of altruism under Darwinian selection. *Theo. Pop. Biol.,* **9**, 360–87.

Matessi, C., and Karlin, S. (1984). On the evolution of altruism by kin selection. *Proc. Natl. Acad. Sci. USA,* **81**, 1754–58.

Matsuoka, S. (1980). Pseudo warning calls in titmice. *Tori,* **29**, 87–90.

May, R. M. (1981). The evolution of cooperation. *Nature,* **292**, 291–92.

May, R. M. (1987). More evolution of cooperation. *Nature,* **327**, 15–16.

Maynard Smith, J. (1964). Group selection and kin selection. *Nature*, **201**, 1145–46.

Maynard Smith, J. (1965). The evolution of alarm calls. *Am. Nat.*, **99**, 59–63.

Maynard Smith, J. (1982). *Evolution and the theory of games.* Cambridge: Cambridge Univ. Press.

Maynard Smith, J., and Price, G. (1973). The logic of animal conflict. *Nature*, **246**, 15–18.

Mepham, B. (1976). *The secretion of milk.* Southhampton: Edward Arnold.

Mesterton-Gibbons, M. (1991). An escape from the Prisoner's Dilemma. *J. Math. Biol.*, **29**, 251–69.

Mesterton-Gibbons, M. (1992a). *An introduction to game-theoretic modelling.* Redwood City, CA: Addison-Wesley.

Mesterton-Gibbons, M. (1992b). On the iterated Prisoner's Dilemma in a finite population. *Bull. Math. Biol.*, **54**, 423–43.

Mesterton-Gibbons, M., and Childress, M. J. (1996). Constraints on reciprocity for non-sessile organisms. *Bull. Math. Biol.*, **58**, 861–875.

Mesterton-Gibbons, M., and Dugatkin, L. A. (1992). Cooperation among unrelated individuals: evolutionary factors. *Q. Rev. Biol.*, **67**, 267–81.

Michelsen, A., Andersen, B. B., Kirchner, W., and Lindauer, M. (1991). The dance language of honeybees—new findings, new perspectives. In *The behaviour and physiology of bees,* (ed. L. J. Goodman and R. C. Fisher), vol. 30, pp. 79–88, New York: CAB International.

Michener, C. D. (1969). Comparative social behavior of bees. *Ann. Rev. Entomol.*, **14**, 299–342.

Michener, C. D. (1974). *The social behavior of bees.* Cambridge: Belknap.

Michener, C. D., and Brothers, D. (1974). Were workers of eusocial hymenoptera initially altruistic or oppressed? *Proc. Natl. Acad. Sci. USA*, **71**, 671–74.

Michener, C. D., and Smith, B. H. (1987). Kin recognition in primitively eusocial insects. In *Kin recognition in animals,* (ed. D. C. Fletcher and C. D. Michener), pp. 209–42, Chichester: Wiley.

Michod, R. (1982). The theory of kin selection. *Ann. Rev. Ecol. System.*, **13**, 23–55.

Michod, R. and Hamilton, W. D. (1980). Coefficients of relatedness in sociobiology. *Nature*, **288**, 694–97.

Michod, R., and Sanderson, M. (1985). Behavioral structure and the evolution of cooperation. In *Evolution—essays in honor of John Maynard Smith* (ed. J. Greenwood and M. Slatkin), pp. 95–104, Cambridge: Cambridge Univ. Press.

Milinski, M. (1977a). Do all members of a swarm suffer the same predation? *Z. Tierpsychol.* **45**, 373–78.

Milinski, M. (1977b). Experiments on the selection by predators against the spatial oddity of their prey. *Z. Tierpsychol.* **43**, 311–25.

Milinski, M. (1987). Tit for Tat and the evolution of cooperation in sticklebacks. *Nature,* **325**, 433–35.

Milinski, M. (1990). No alternative to Tit for Tat in sticklebacks. *Anim. Behav.,* **39**, 989–91.

Milinski, M. (1992). Predator inspection: cooperation or "safety in numbers"? *Anim. Behav.,* **43**, 679–81.

Milinski, M. (1996). By-product mutualism, Tit-for-Tat reciprocity, and cooperative predator inspection: a reply to Connor. *Anim. Behav.,* **51**, 458–461.

Milinski, M., and Bolthauser, P. (1995). Boldness and predator deterrence: a critique to Godin and Davis. *Proc. R. Soc. Lond, Ser. B*, **262**, 103–5.

Milinski, M., Kulling, D., and Kettler, R. (1990a). Tit for Tat: sticklebacks "trusting" a cooperating partner. *Behav. Ecol.,* **1,** 7–12.

Milinski, M., Pfugler, D., Kulling, D., and Kettler, R. (1990b). Do sticklebacks cooperate repeatedly in reciprocal pairs? *Behav. Ecol. Sociobiol.,* **27,** 17–23.

Millar, J. S. (1977). Adaptive features of mammalian reproduction. *Evol.,* **31,** 370–86.

Mitman, G. (1992). *The state of nature: ecology, community and American social thought, 1900–1950.* Chicago: Univ. of Chicago Press.

Mittlebach, G. (1984). Group size and feeding rate in bluegills. *Copeia,* **1984,** 998–1000.

Mirmirani, M., and Oster, G. (1978). Competition, kin selection and evolutionary stable strateties. *Theo. Pop. Biol.,* **13,** 304–39.

Mock, D. (1980). White-dark polymorphism in herons. In *Proceedings of the first Welder Wildlife Symposium,* (ed. D. L. Drawe), pp. 145–161, Sinton, TX: Welder Wildlife Foundation.

Moehlman, P. (1979). Jackal helpers and pup survival. *Nature,* **277,** 382–83.

Moehlman, P. (1989). Interspecific variation in canid social systems. In *Carnivore behavior, ecology and evolution,* (ed. J. L. Gittleman), pp. 143–163, Ithaca: Cornell Univ. Press.

Moller, A. P. (1988). False alarm calls as a means of resource usurpation in the great tit *Parus major. Ethology,* **79,** 25–30.

Moller, A. P. (1990). Deceptive use of alarm calls by male swallows, *Hirundo rustica:* a new paternity guard. *Behav. Ecol.,* **1,** 1–6.

Montagu, A. (1952). *Darwin: competition and cooperation.* New York: H Schuman.

Montevecchi, W. A. (1979). Predator-prey interactions between ravens and kittiwakes. *Z. Tierpsychol.,* **49,** 136–41.

Montgomery, W. L., Gerrodette, T., and Marshall, L. D. (1980). Response of algal communities to grazing by the yellowtail surgeonfish *Prionurus punctatus* in the Gulf of California, Mexico. *Bull. Mar. Sci.,* **30,** 477–81.

Moore, A. J., Breed, M. D., and Moore, M. J. (1987). Characterization of guard behavior in honeybees, *Apis mellifera. Anim. Behav.,* **35,** 1159–67.

Moore, J. (1984). The evolution of reciprocal sharing. *Ethol. Sociobiol.,* **5,** 5–14.

Mooring, M. S., and Hart, B. L. (1992). Reciprocal allogrooming in dam-reared and hand-reared impala fawns. *Ethology,* **90,** 37–51.

Mooring, M. S., and Hart, B. L. (1995). Costs of allogrooming in impala: distraction from vigilance. *Anim. Behav.,* **49,** 1414–16.

Morse, D. H. (1970). Ecological aspects of some mixed species foraging flocks of birds. *Ecol. Monographs,* **4,** 119–68.

Morse, P. M. (1958). *Queues, inventories and maintenance.* New York: Wiley and Sons.

Morin, P. A., Moore, J. J., Chakraborty, R., Jin, L., Goodall, J., and Woodruff, D. S. (1994). Kin selection, social structure, gene flow, and the evolution of chimpanzees. *Science,* **265,** 1193–1201.

Morton, E. S. (1977). On the occurrence and significance of motivation-structural rules in some bird and animal sounds. *Am. Nat.,* **111,** 855–69.

Moss, C. (1976). *Portraits in the wild: behavior studies of East African mammals.* Boston: Houghton Mifflin.

Motta, P. (1983). Response by potential prey to coral reef predators. *Anim. Behav.,* **31,** 1257–59.

Motro, U. (1991). Cooperation and defection: playing the field and the ESS. *J. Theo. Biol.,* **151,** 145–54.

Moynihan, M. (1970). Some behavior patterns of platyrrhine monkeys. II. *Saginus geoffoyi* and some other tamarins. *Smithson. Contr. Zool.*, **28**.

Mueller, L. D., and Feldman, M. W. (1988). The evolution of altruism by kin selection: new phenomena with strong selection. *Ethol. Sociobiol.*, **9**, 223–39.

Munn, C. A. (1986). Birds that "cry wolf." *Nature*, **319**, 143–45.

Murray, M. G. (1981). Structure of association in impala, *Aepyceros melampus*. *Behav. Ecol. Sociobiol.*, **9**, 23–33.

Murray, M. G. (1982). Home range, disposal and the clan system of impala. *Afr. J. Ecol.*, **20**, 253–69.

Newman, J., and Caraco, T. (1989). Co-operative and noncooperative bases of food-calling. *J. Theo. Biol.*, **141**, 197–209.

Nicolson, N. A. (1987). Infants, mothers and other females. In *Primate societies*, (ed. B. Smuts, D. L. Cheney, R. Seyfarth, W. R., and T. Struhsaker), pp. 330–342, Chicago: Univ. of Chicago Press.

Nishida, T. (1988). Development of social grooming between mother and offspring in wild chimpanzees. *Folia Primatol.*, **50**, 109–23.

Nishida, T., Hasegawa, T., Hayaki, H., Takahata, Y., and Uehara, S. (1992). Meat-sharing as a coalition strategy by an alpha male chimpanzee? Topics in Primatology **1**, 159–74.

Nishida, T., Hiraiwa-Hasegawa, M., Hasegawa, T., and Takahata, Y. (1985). Group extinction and female transfer in wild chimpanzees in the Mahale National Park, Tanzania. *Z. Tierpsychol.*, **67**, 284–301.

Nishida, T., Uehara, S., and Nyondo, R. (1983). Predatory behaviour among wild chimpanzees of the Mahale Mountains. *Primates*, **20**, 1–20.

Noë, R. (1986). Lasting alliances among adult male Savannah baboons. In *Primate ontogeny, cognition and social behaviour*, (ed. J. Else and P. Lee), pp. 381–92, Cambridge: Cambridge Univ. Press.

Noë, R. (1990). A veto game played by baboons: a challenge to the Prisoner's Dilemma as a paradigm for reciprocity and cooperation. *Anim. Behav.*, **39**, 78–90.

Noë, R. (1992). Alliance formation among male baboons: shopping for profitable partners. In *Coalitions and alliances in humans and other animals* (ed. A. Harcourt and F. de Waal), pp. 285–321, Oxford: Oxford Univ. Press.

Noë, R., van Schaik, C., and van Hooff, J. (1991). The market effect: an explanation for pay-off asymmetries among collaborating animals. *Ethology*, **87**, 97–118.

Nonacs, P. (1988). Queen number in colonies of social hymenoptera as a kin-selected adaptation. *Evol.*, **42**, 566–80.

Nonacs, P. (1993). The economics of brood raiding and nest consolidation during ant colony founding. *Evol. Ecol.*, **7**, 625–33.

Norris, K., and Schilt, C. R. (1988). Cooperative societies in three-dimensional space: on the origins of aggregations, flocks and schools, with special reference to dolphins and fish. *Ethol. Sociobiol.*, **9**, 149–79.

Norris, K. S. (1958). The big one that got away. *Pac. Discovery*, **11**, 3–9.

Nowak, M. (1990a). An evolutionarily stable strategy may be inaccessible. *J. Theo. Biol.*, **142**, 237–41.

Nowak, M. (1990b). Stochastic strategies in the Prisoner's Dilemma. *Theo. Pop. Biol.*, **38**, 93–112.

Nowak, M., and Sigmund, K. (1989). Oscillations in the evolution of reciprocity. *J. Theo. Biol.*, **137**, 21–26.

Nowak, M., and Sigmund, K. (1990). The evolution of stochastic strategies in the Prisoner's Dilemma. *Acta. Appl. Math.*, **20**, 247–65.

Nowak, M., and May, R. M. (1992). Evolutionary games and spatial chaos. *Nature*, **359**, 826–29.

Nowak, M., and Sigmund, K. (1992). Tit for Tat in heterogeneous populations. *Nature*, **355**, 250–252.

Nowak, M., and Sigmund, K. (1993a). A strategy of win-stay, lose-shift that outperforms tit-for-tat in the Prisoner's Dilemma game. *Nature*, **364**, 56–58.

Nowak, M., and Sigmund, K. (1993b). Chaos and the evolution of cooperation. *Proc. Natl. Acad. Sci. USA*, **90**, 5091–94.

Nowak, M., and Sigmund, K. (1994). The alternating Prisoner's Dilemma. *J. Theo. Biol.*, **168**, 219–26.

Nunney, L. (1985). Group selection, altruism, and structured deme models. *Am. Nat.*, **126**, 212–30.

O'Riain, M. J., Jarvis, J. U., and Faulkes, C. G. (1996). A dispersive morph in the naked mole-rat. *Nature*, **380**, 619–621.

Ohguchi, O. (1981). Prey density and selection against oddity by three- spined stickle-backs. *Z. Tierpsychol.*, Supplement #23.

Oki, J., and Maeda, Y. (1973). Grooming as a regulator of behaviour in Japanese macaques. In *Behavioral regulators of behavior in primates*, (ed. R. C. Carpenter), pp. 149–163, Lewisburg, PA: Bucknell Univ. Press.

Oster, G. F., Eshel, I., and Cohen, D. (1977). Worker-queen conflict and the evolution of social insects. *Theo. Pop. Biol.*, **12**, 49–85.

Oster, G. F., and Wilson, E. O. (1978). *Caste and ecology in the social insects.* Princeton, NJ: Princeton Univ. Press.

Owens, D. D., and Owens, M. J. (1979). Communal denning and clan associations in brown hyenas *(Hyaena brunnea, Thunberg)* of the central Kalahari Desert. *Afr. J. Ecol.*, **17**, 35–44.

Owens, D. D., and Owens, M. J. (1984). Helping behavior in brown hyenas. *Nature*, **308**, 843–45.

Packer, C. (1975). Male transfer in olive baboons. *Nature*, **255**, 219–20.

Packer, C. (1977). Reciprocal altruism in *Papio anubis. Nature*, **265**, 441–43.

Packer, C. (1979). Male dominance and reproductive activity in *Papio anubis. Anim. Behav.*, **27**, 37–45.

Packer, C. (1980). Male care and exploitation of infants in *Papio anubis. Anim. Behav.*, **28**, 512–20.

Packer, C., and Abrams, P. (1990). Should co-operative groups be more vigilant than selfish groups? *J. Theo. Biol.*, **142**, 341–57.

Packer, C., Collins, D. A., Sindimwo, A., and Goodall, J. (1994). Reproductive constraints on aggressive competition in female baboons. *Nature*, **373**, 60–63.

Packer, C., Gilbert, D., Pusey, A. E., and O'Brien, S. (1991). A molecular genetic analysis of kinship and cooperation in African lions. *Nature*, **351**, 562–65.

Packer, C., Herbst, L., Pusey, A. E., Bygott, J. D., Hanby, J. P., Cairns, S. J., and Borgerhoff-Mulder, M. (1988). Reproductive success of lions. In *Reproductive success*, (ed. T. H. Clutton-Brock), pp. 363–83, Chicago: Univ. of Chicago Press.

Packer, C., Lewis, S., and Pusey, A. E (1992). A comparative analysis of non-offspring nursing. *Anim. Behav.*, **43**, 265–81.

Packer, C., and Pusey, A. E. (1982). Cooperation and competition within coalitions of male lions: kin selection or game theory? *Nature*, **296**, 740–42.

Packer, C., and Ruttan, L. (1988). The evolution of cooperative hunting. *Am. Nat.*, **132**, 159–94.

Page, R. E. (1986). Sperm utilization in social insects. *Ann. Rev. Entomol.*, **31**, 297–320.

Page, R. E., and Breed, M. D. (1987). Kin recognition in social bees. *Trends Ecol. Evol.*, **2**, 272–75.

Page, R. E., and Fondrk, M. K. (1995). The effects of colony-level selection on the social organization of honey bee *(Apis Mellifera L.)* colonies: colony-level components of pollen hoarding. *Behav. Ecol. Sociobiol.*, **36**, 135–44.

Page, R. E., Waddington, K. D., Hunt, G. J., and Fondrk, M. K. (1995). Genetic determinants of honeybee foraging behavior. *Anim. Behav.*, **50**, 1617–25.

Pamilo, P. (1982). Genetic evolution of sex ratios in eusocial hymenoptera: allele frequency simulations. *Am. Nat.*, **119**, 638–56.

Parker, G., and Rubenstein, D. (1981). Role assessment, reserve strategy and the acquisition of information in asymmetric animal contests. *Anim. Behav.*, **29**, 221–40.

Parker, P. G., Waite, T. A., Heinrich, B., and Marzluff, J. M. (1994). Do common ravens share ephemeral food resources with kin? DNA fingerprinting evidence. *Anim. Behav.*, **48**, 1085–93.

Partridge, B. L., Johansson, J., and Kalish, J. (1983). The structure of schools of giant bluefin tuna in Cape Cod bay. *Env. Biol. Fishes*, **9**, 253–62.

Patten, C. J. (1920). *The grand strategy of evolution.* Boston: Badger.

Patten, C. J. (1925). *The passing of phantoms: a study of evolutionary psychology and morals.* New York: E. P. Dutton.

Pearl, R. (1939). The evolution of sociality. *Ecology,* **20**, 305–10.

Peck, J. R., and Feldman, M. (1986). The evolution of helping behavior in a large randomly mixed population. *Am. Nat.*, **127**, 209–21.

Peck, J. R. (1993). Friendship and the evolution of cooperation. *J. Theo. Biol.*, **162**, 195–228.

Pepper, J. W., Braude, S. H., Lacey, E. A., and Sherman, P. W. (1991). Vocalizations of the naked mole-rat. In *The biology of the naked mole-rat,* (ed. P. W. Sherman, J. Jarvis, and R. D. Alexander), pp. 243–74, Princeton, NJ: Princeton Univ. Press.

Pereira, M. E., and Izard, M. K. (1989). Lactation and care for unrelated infants in forest-living ringtailed lemurs. *J. Primatol.*, **18**, 101–18.

Perrins, C. (1968). The purpose of the high-intensity alarm call in small passerines. *Ibis*, **110**, 200–201.

Pettifor, R. A. (1990). The effects of avian mobbing on a potential predator, the European kestrel, *Falco tinnunculus. Anim. Behav.*, **39**, 821–27.

Petersen, C. W. (1987). Reproductive behavior and gender allocation in *Serranus fasciatus,* a hermaphroditic reef fish. *Anim. Behav.*, **35**, 1601–14.

Petersen, C. W. (1990). The relationships among population density, individual size, mating tactics and reproductive success in a hermaphroditic reef fish, *Serranus fasciatus. Behaviour,* **113**, 57–80.

Petersen, C. W. (1991). Sex allocation in hermaphorditic sea basses. *Am. Nat.*, **138**, 650–67.

Petersen, C. W. (1995). Reproductive behavior, egg trading and correlates of male mating success in the simulataneous hermaphrosite, *Serranus tabacarius. Env. Biol. Fishes,* **43**, 351–61.

Pfeiffer, W. (1977). The distribution of fright reaction and alarm cells in fishes. *Copeia,* **1977**, 653–65.

Pfennig, D. W. (1995). Absence of joint nesting advantage in desert seed harvester ants: evidence from a field experiment. *Anim. Behav.*, **49**, 567–75.

Pilleri, G., and Knuckey, J. (1969). Behavior patterns of some delphionidae observed in the western Mediterranean. *Z. Tierpsychol.*, **26**, 48–72.

Pitcher, T. (ed.) (1986). *The behavior of teleost fishes.* Baltimore, MD: Johns Hopkins Univ. Press.

Pitcher, T. (1993). Who dares wins: the function and evolution of predator inspection behaviour in shoaling fish. *Neth. J. Zool.,* **42,** 371–91.

Pitcher, T., Green, D. A., and Magurran, A. E. (1986). Dicing with death: predator inspection behavior in minnow shoals. *J. Fish Biol.,* **28,** 439–48.

Pollock, G. (1988). Population structure, spite and the iterated Prisoner's Dilemma. *Am. Jour. Phys. Anthro.,* **77,** 209–21.

Pollock, G. (1989). Evolutionary stability of reciprocity in a viscous lattice. *Soc. Net.,* **11,** 175–213.

Pollock, G., and Dugatkin, L. A. (1992). Reciprocity and the evolution of reputation. *J. Theo. Biol.,* **159,** 25–37.

Pollock, G., and Rissing, S. (1984). Mating season and colony foundation of the seed harvester ant, *Veromessor pergandei. Psyche,* **92,** 125–34.

Poppleton, F. (1957). The birth of an elephant. *Oryx,* **4,** 180–81.

Potts, G. W. (1983). The predatory tactics of *Caranx melampygus* and the response of its prey. In *Predators and prey in fishes* (ed. D. L. G. Noakes, B. G. Lindquist, G. S. Helfman, and J. A. Ward), pp. 181–91, The Hague: Junk.

Poulin, R. (1993). A cleaner perspective on cleaning symbiosis. *Rev. Fish. Biol. Fisheries,* **3,** 75–79.

Poulin, R., and Vickery, W. L. (1995). Cleaning symbiosis as an evolutionary game: to cheat or not to cheat? *J. Theo. Biol.,* **175,** 63–70.

Poundstone, W. (1992). *Prisoner's dilemma: Jon Von Neuman, game theory and the puzzle of the bomb.* New York: Doubleday.

Pratt, H. M. (1980). Directions and timing of great blue heron foraging flights from a California colony: implications for social facilitation of food finding. *Wilson Bull.,* **92,** 489–96.

Price, G. R. (1972). Extension of covariance selection mathematics. *Ann. Hum. Genet.,* **35,** 485–90.

Pulliam, R. (1973). On the advantages of flocking. *J. Theo. Biol.,* **38,** 419–22.

Queller, D. C. (1985). Kinship, reciprocity and synergism in the evolution of social behavior. *Nature,* **318,** 366–67.

Queller, D. C. (1989). The evolution of eusociality: reproductive head starts of workers. *Proc. Natl. Acad. Sci. USA,* **86,** 3224–26.

Queller, D. C. (1992). Quantitative genetics, inclusive fitness and group selection. *Amer. Nat.,* **139,** 540–58.

Queller, D. C. (1996). The measurement and meaning of inclusive fitness. *Anim. Behav.,* **51,** 229–32.

Queller, D. C., Negron-Sotomayor, J. A., Strassmann, J., and Hughes, C. (1993). Queen number and genetic relatedness in a neotropical wasp, *Polybia occidentalis. Behav. Ecol.,* **4,** 7–13.

Queller, D. C., and Strassman, J. (1988). Reproductive success and group nesting in the paper wasp, *Polistes annularis.* In *Reproductive success,* (ed. T. Clutton-Brock), pp. 76–96, Chicago: Univ. of Chicago Press.

Queller, D. C., and Strassman, J. (1989). Measuring inclusive fitness in social wasps. In *The genetics of social behavior,* (ed. M. Breed and R. Page), pp. 103–22, Boulder: Westview Press.

Queller, D. C., Strassmann, J., and Hughes, C. (1988). Genetic relatedness in colonies of tropical wasps with multiple queens. *Science,* **242,** 1155–57.

Queller, D. C., Strassman, J., and Hughes, C. (1992). Genetic relatedness and population structure in primitively eusocial wasps in the genus *Mischocyttarus. J. Hym. Res.,* **1,** 81–89.

Rabenold, P. (1987). Recruitment to food in black vultures: evidence for following from communal roosts. *Anim. Behav.,* **35,** 1775–85.

Radner, D., and Radner, M. (1989). *Animal consciousness.* Buffalo, NY: Prometheus Books.

Ranta, E., and Lindstrom, K. (1990). Assortative schooling in three-spined sticklebacks? *Ann. Zool. Fennici,* **27,** 67–75.

Ranta, E., Lindstrom, K., and Peuhkuri, N. (1992). Size matters when three-spined sticklebacks go to school. *Anim. Behav.,* **43,** 160–62.

Rasa, O. A. (1977). The ethology and sociology of the dwarf mongoose (*Helogale undulata rufala*). *Z. Tierpsychol.,* **43,** 337–405.

Rasa, O. A. (1987). The dwarf mongoose: a study of the behavior and social structure in relation to ecology of a small social carnivore. *Adv. Study Behav.,* **17,** 121–63.

Ratnieks, F. L. (1995). Evidence for queen-produced egg-marking pheromone and its use in worker policing in the honeybee. *J. Apiculture Res.,* **34,** 31–37.

Ratnieks, F. L., and Visscher, P. K. (1988). Reproductive harmony via mutual policing by workers in eusocial hymenoptera. *Am. Nat.,* **132,** 217–36.

Ratnieks, F. L., and Visscher, P. K. (1989). Worker policing in the honeybee. *Nature,* **342,** 796–97.

Reboreda, J. C. and Kacelnik, A. (1990). On cooperation, tit for tat and mirrors. *Anim. Behav.,* **40,** 1188–89.

Reeve, H. K. (1991). Polistes. In *The biology of social wasps,* (ed. K. G. Ross and R. W. Matthews), pp. 99–148, Ithaca: Cornell Univ. Press.

Reeve, H. K. (1992). Queen activation of lazy workers in colonies of the eusocial naked mole rat. *Nature,* **358,** 147–49.

Reeve, H. K. (1993). Haplodiploidy, eusociality and the absence of male alloparental care in Hymenoptera: a unifying genetic hypothesis distinct from kin selection theory. *Phil. Trans. R. Soc. Lond. B,* **342,** 335–52.

Reeve, H. K., and Dugatkin, L. A. (1996). Protected invasion and cooperation between unrelated individuals. Manuscript.

Reeve, H. K., and Nonacs, P. (1992). Social contracts in wasp societies. *Nature,* **359,** 823–25.

Reeve, H. K., and Ratnieks, F. L. (1993). Queen-queen conflicts in polygynous societies: mutual tolerance and reproductive skew. In *Queen number and sociality in insects,* (ed. L. Keller), pp. 45–86, Oxford: Oxford Univ. Press.

Reeve, H. K., and Shellman-Reeve, J. (in review). Sex biases in parental and alloparental care: the general protected invasion theory.

Reeve, H. K., and Sherman, P.W. (1991). Intracolonial aggression and nepotism by the breeding female naked mole rat. In *The biology of the naked mole-rat,* (ed. P. W. Sherman, J. Jarvis, and R. D. Alexander), pp. 337–357, Princeton, NJ: Princeton Univ. Press.

Reeve, H. K., Westneat, D. F., Noon, W. A., Sherman, P. W., and Aquadro, C. F. (1990). DNA "fingerprinting" reveals high levels of inbreeding in colonies of the eusocial naked mole rat. *Proc. Natl. Acad. Sci. USA,* **87,** 2496–500.

Regelmann, K., and Curio, E. (1986). How do great tit *(Parus major)* pair mates cooperate in brood defence? *Behaviour,* **97,** 10–36.

Reidman, M. L. (1982). The evolution of alloparental care and adoption in mammals and birds. *Q. Rev. Biol.,* **57,** 405–35.

Reinhardt, V., and Reinhardt, A. (1980). Cohesive relationships in a cattle *(Bos indicus)* herd. *Behaviour*, **75**, 120–51.

Reinheimer, H. (1913). *Evolution by cooperation, a study in bioeconomics.* London: Kegan Paul, Trench and Trubner.

Reinheimer, H. (1920). *Symbiosis, a sociophysiological study of evolution.* London: Headley Brothers.

Rhine, R. J. (1973). Variations and consistency in the social behavior of groups of stump-tailed macaques. *Primates*, **14**, 21–35.

Rhine, R. J. (1981). Adult positioning in baboon progressions: order and chaos revisited. *Folia Primatol.*, **35**, 77–116.

Rijkesen, H. (1978). Hunting behavior in hominids: some ethological aspects. In *Recent advances in primatology,* vol. 3: *Evolution* (ed. D. Chivers and K. Joysey), pp. 499–502, London: Academic Press.

Rinderer, T. E. (1982). Regulated nectar harvesting by the honeybee. *J. Apic. Res.*, **21**, 74–87.

Rissing, S., and Pollock, G. (1986). Social interaction among pleometric queens of *Veromessor pergandei* during colony foundation. *Anim. Behav.*, **34**, 226–34.

Rissing, S., and Pollock, G. (1987). Queen aggression, pleometric advantage and brood raiding in the ant *Veromessor pergandei. Anim. Behav.*, **35**, 975–82.

Rissing, S., and Pollock, G. (1988). Pleometrosis and polygyny in ants. In *Interindividual behavioral variability in social insects,* (ed. R. Jeanne), pp. 170–222, New York: Westview Press.

Rissing, S., and Pollock, G. (1991). An experimental analysis of pleometric advantage in *Messor pergandei. Ins. Soc.*, **63**, 205–11.

Rissing, S., Pollock, G., Higgins, M., Hagen, R., and Smith, D. (1989). Foraging specialization without relatedness or dominance among co-founding ant queens. *Nature*, **338**, 420–22.

Robertson, D. R., Sweatman, E. A., Fletcher, E. A., and Cleland, M. E. (1976). Schooling as a mechanism for circumventing the territory of competitors. *Ecology*, **57**, 1208–20.

Robins, C. R., and Starck, W. A. (1961). Materials for a revision of Serranus and related genera. *Proc. Natl. Acad. Sci., Phila.*, **113**, 259–314.

Robinson, G. E., and Page, R. E. (1988). Genetic determination of guarding and undertaking in honey-bee colonies. *Nature*, **333**, 356–58.

Roell, A. (1978). Social behaviour of the jackdaw, *Corvus monedula,* in relation to its niche. *Behaviour*, **64**, 1–124.

Rohwer, S. (1978). Parental cannibalism of offspring and egg raiding as a courtship strategy. *Am. Nat.*, **112**, 429–40.

Romanes, G. J. (1895). *Mental evolution in animals.* New York: Appelton.

Romanes, G. J. (1898). *Animal intelligence.* London: Kegan Paul, Trench, and Trubner.

Rood, J. P. (1974). Banded mongoose males guard young. *Nature*, **248**, 176.

Rood, J. P. (1978). Dwarf mongoose helpers at the den. *Z. Tierpsychol.*, **48**, 227–87.

Rood, J. P. (1983). The social system of the dwarf mongoose. In *Recent advances in the study of mammalian behavior* (ed. J. F. Eisenberg and D. G. Kleiman), pp. 454–88, Shippensburg, PA: American Society of Mammalogists.

Rood, J. P. (1986). Ecology and social evolution in the mongooses. In *Ecological aspects of social evolution* (ed. D. I. Rubenstein and R. W. Wrangham), pp. 131–52, Princeton: Princeton Univ. Press.

Rood, J. P. (1990). Group size, survival, reproduction and routes to breeding in dwarf mongooses. *Anim. Behav.*, **39**, 566–72.

Rosenblum, L. A., Kaufman, I., and Stynes, I. C. (1966). Some characteristics of adult social autogrooming in two species of macaques. *Folia Primatol.*, **4**, 438–51.

Ross, K. G. (1986). Kin selection and the problem of sperm utilization in social insects. *Nature*, **323**, 798–800.

Ross, K. G., and Carpenter, J. M. (1991). Population genetic structure, relatedness, and breeding systems. In *The social biology of wasps* (ed. K. G. Ross and R. W. Matthews), pp. 451–79, Ithaca, NY: Cornell Univ. Press.

Rothstein, S., and Pirotti, R. (1987). Distinctions among reciprocal altruism, kin selection and cooperation and a model for the initial evolution of beneficent behavior. *Ethol. Sociobiol.*, **9**, 189–209.

Rowell, T. E., Hinde, R. A., and Spencer-Booth, Y. (1964). "Aunt"-infant interaction in captive rhesus monkeys. *Anim. Behav.*, **12**, 219–26.

Russell, J. K. (1983). Altruism in coati bands: nepotism or reciprocity? In *Social behavior of female vertebrates,* (ed. S. Wasser), pp. 263–90, New York: Academic Press.

Ryti, R. (1988). Geographic variation in cooperative colony foundation in *Veromessor pergandei. Pan-Pacific Entomologist*, **63**, 255–57.

Ryti, R. T., and Case, T. J. (1984). Spatial arrangement and diet overlap between colonies of desert ants. *Oecologia*, **62**, 401–4.

Sade, D. S. (1972). Sociometrics of *Macaca mulatta*. I. Linkages and cliques of free-ranging baboons *(Papio cyanocephalus ursinus). Folia Primatol.*, **18**, 196–223.

Sade, D. S., Cushing, K., Cushing, C., Dunaif, J., Figueroa, A., Kaplan, J. R., Lauer, C., Rhodes, D., and Schneider, J. (1976). Population dynamics in relation to social structure on Cayo Santiago. *Yrbk. Phys. Anthr.*, **20**, 253–62.

Saunders, C. D. (1987). Grooming quality in relation to tick density and reciprocity between partners. *Am. J. Primatology*, **12**, 369.

Savage-Rumbaugh, S., Rumbaugh, D., and Boysen, S. (1978). Linguistically mediated tool use and exchange by chimpanzees *(Pan troglodytes). Behav. Brain Sci.*, **4**, 539–54.

Schaller, G. B. (1972). *The Serengeti lion: a study of predator-prey relations.* Chicago: Univ. of Chicago Press.

Schmid-Hempel, P. (1986). Do honeybees get tired? The effect of work load on patch departure. *Anim. Behav.*, **34**, 1243–50.

Schmid-Hempel, P. (1987). Efficient nectar collection by honeybees. I. Economic models. *J. Anim. Ecol.*, **56**, 209–18.

Schmid-Hempel, P. (1990). Reproductive competition and the evolution of work load in social insects. *Am. Nat.*, **135**, 501–26.

Schmid-Hempel, P. (1991). The ergonomics of worker behavior in social hymenoptera. *Adv. Study. Behav.*, **20**, 87–133.

Schmid-Hempel, P., Kacelnik, A., and Houston, A. (1985). Honeybees maximize efficiency by not filling their crop. *Behav. Ecol. Sociobiol.*, **17**, 61–66.

Schmid-Hempel, P., Winston, M., and Ydenberg, R. C. (1993). Foraging of individual workers in relation to colony state in the social hymenoptera. *Can. Entomol.*, **125**, 129–60.

Scheel, D., and Packer, C. (1991). Group hunting behaviour of lions: a search for cooperation. *Anim. Behav.*, **41**, 697–709.

Schino, G., Scuccio, S., Maestrippi, D., and Turillazzi, P. G. (1988). Allogrooming as a tension reduction mechanism: a behavioral approach. *Am. J. Primatology*, **16**, 43–50.

Schmitt, R. J., and Strand, S. W. (1982). Cooperative foraging by yellowtail *Seriola lalandei* (Carangidae) on two species of fish prey. *Copeia,* **1982,** 714–17.

Schwagmeyer, T. (1980). Alarm calling of the thirteen-lined ground squirrel. *Behav. Ecol. Sociobiol,* **7,** 195–200.

Scott, J. P. (1958). *Animal behavior.* Chicago: Univ. of Chicago Press.

Seeley, T. D. (1983). Division of labor between scouts and recruits in honeybee foraging. *Behav. Ecol. Sociobiol.,* **12,** 253–59.

Seeley, T. (1985). *Honeybee ecology: a study of adaptation in social life.* Princeton: Princeton Univ. Press.

Seeley, T. D. (1986). Social foraging by honeybees: how colonies allocate foragers among patches of flowers. *Behav. Ecol. Sociobiol.,* **19,** 343–54.

Seeley, T. (1989). The honeybee colony as a superorganism. *Am. Sci.,* **77,** 546–53.

Seeley, T. (1991). Collective decision-making in honeybees: how colonies choose among nectar sources. *Behav. Ecol. Sociobiol.,* **28,** 277–90.

Seeley, T. (1992). The tremble dance of the honeybee: message and meanings. *Behav. Ecol. Sociobiol.,* **31,** 375–83.

Seeley, T. (1995). *The wisdom of the hive.* Cambridge, MA: Harvard Univ. Press.

Seeley, T. D., and Levien, R. A. (1987). Social foraging by honeybees: how a colony tracks rich sources of nectar. In *Neurobiology and behavior in honeybees* (ed. R. Menzel and A. Mercer), pp. 38–52. Berlin: Springer-Verlag.

Seeley, T. D., and Seeley, R. H. (1982). Colony defense strategies of the honeybees in Thailand. *Ecology,* **52,** 43–63.

Seeley, T. D., and Towne, W. F. (1989). Social foraging in honeybees: how nectar foragers assess their colony's nutritional status. *Behav. Ecol. Sociobiol.,* **24,** 181–99.

Seeley, T. D., and Towne, W. F. (1992). Tactics of dance choice in honeybees: do foragers compare dances? *Behav. Ecol. Sociobiol.,* **30,** 59–69.

Seger, J. (1981). Kinship and covariance. *J. Theo. Biol.,* **91,** 191–213.

Seger, J. (1983). Partial voltinism may cause sex-ratio biases that favor eusociality. *Nature,* **301,** 59–62.

Seger, J. (1989a). Who are the drone police? *Nature,* **342,** 741–42.

Seger, J. (1989b). All for one, one for all, that is our device. *Nature,* **338,** 374–75.

Seger, J. (1991). Cooperation and conflict in social insects. In *Behavioural ecology:* 3rd ed. (ed. J. R. Krebs and N. B. Davies), pp. 338–73, Oxford: Blackwell.

Sella, G. (1985). Reciprocal egg trading and brood care in a hermaphroditic polycheate worm. *Anim. Behav.,* **33,** 938–44.

Sella, G. (1988). Reciprocation, reproductive success and safeguards against cheating in a hermaphroditic polycheate worm. *Biol. Bull.,* **175,** 212–17.

Sella, G. (1991). Evolution of biparental care in the hermaphroditic polychaete worm *Ophryotrocha diadema.* *Evol.,* **45,** 63–68.

Selten, R., and Hammerstein, P. (1984). Gaps in Harley's argument on the evolutionarily stable learning rules and the logic of "tit for tat". *Behav. Brain Sci.,* **7,** 115–16.

Seyfarth, R. M. (1976). Social relationships among adult female baboons. *Anim. Behav.,* **24,** 917–38.

Seyfarth, R. M. (1977). A model of social grooming among adult female primates. *J. Theo. Biol.,* **65,** 671–98.

Seyfarth, R. M. (1980). The distribution of grooming and related behaviours among adult female vervet monkeys. *Anim. Behav.,* **28,** 798–813.

Seyfarth, R. M., and Cheney, D. L. (1984). Grooming alliances and reciprocal altruism in vervet monkeys. *Nature,* **308,** 541–43.

Seyfarth, R. M., and Cheney, D. L. (1988). Empirical tests of reciprocity theory: problems in assessment. *Ethol. Sociobiol.*, **9**, 181–87.

Seyfarth, R. M., Cheney, D. L., and Marler, P. (1980a). Monkey responses to three different alarm calls: evidence for predator classification and semantic communication. *Science*, **210**, 801–3.

Seyfarth, R. M., Cheney, D. L., and Marler, P. (1980b). Vervet monkey alarm calls: semantic communication in a free-ranging primate. *Anim. Behav.*, **28**, 1070–94.

Shedd, D. H. (1982). Seasonal variation and function of mobbing and related antipredator behaviors of the American robin *(Turdus migratorius)*. *Auk*, **99**, 342–46.

Shedd, D. H. (1983). Seasonal variation in mobbing intensity in the black-capped chickadee. *Wilson Bull.*, **95**, 343–48.

Sherman, P. W. (1977). Nepotism and the evolution of alarm calls. *Science*, **197**, 1246–53.

Sherman, P. W. (1980). The limits of ground squirrel nepotism. In *Sociobiology: beyond nature/nurture?* (ed. G. Barlow and J. Silverberg), pp. 505–44, Boulder, CO: Westview.

Sherman, P. W. (1981). Kinship, demography, and Belding's ground squirrel nepotism. *Behav. Ecol. Sociobiol.*, **8**, 251–59.

Sherman, P. W. (1985). Alarm calls of Belding's ground squirrels to aerial predators: nepotism or self-preservation? *Behav. Ecol. Sociobiol.*, **17**, 313–23.

Sherman, P. W., Jarvis, J., and Alexander, R. (eds.) (1991). *The biology of the naked mole-rat*. Princeton, NJ: Princeton Univ. Press.

Sherman, P. W., Jarvis, J., and Braude, S. H. (1992). Naked mole rats. *Sci. Am.* **267**, 72–9.

Sherman, P. W., Lacey, E. A., Reeve, H. K., and Keller, L. (1995). The eusociality continuum. *Behav. Ecol.*, **6**, 102–8.

Siebenaler, J. B., and Caldwell, D. K. (1956). Cooperation among adult dolphins. *J. Mammal.*, **37**, 126–28.

Sikkel, P. (1986). Intraspecific cleaning by juvenile salema, *Xenestius californiensis*. *Calif. Fish and Game*, **72**, 170–72.

Silk, J. (1980). Kidnapping and female competition among captive bonnet macaques. *Primates*, **21**, 100–110.

Silk, J. (1982). Altruism among female *Macaca radiata:* explanations and analysis of patterns of grooming and coalition formation. *Behaviour*, **79**, 162–88.

Silk, J. (1992a). The patterning of intervention among male bonnet macaques: reciprocity, revenge and loyalty. *Current Anthropology*, **33**, 319–24.

Silk, J. (1992b). Patterns of intervention in agonistic contests among male bonnet macaques. In *Coalitions and alliances in humans and other animals*, (ed. A. Harcourt and F. B. M. de Waal), pp. 214–32, Oxford: Oxford Univ. Press.

Silk, J., Clark-Wheatley, C., Rodman, P. S., and Samuels, A. (1981). Differential reproductive success and facultative adjustment of sex ratios among captive female bonnet macaques (*Macaca radiata*). *Anim. Behav.*, **29**, 1106–20.

Silk, J., Rodman, P. S., and Samules, A. (1980). Rank, reproductive success and skewed sex ratios in *Macaca radiata*. *Amer. J. Phys. Anthropol.*, **52**, 279.

Silk, J., Samuels, A., and Rodman, P. (1981). The influence of kinship, rank, and sex on affiliation and aggression between adult female and immature bonnet macaques (*Macaca radiata*). *Behaviour*, **78**, 111–33.

Skutch, A. F. (1935). Helpers at the nest. *Auk*, **52**, 257–73.

Skutch, A. F. (1987). *Helpers at bird's nests*. Iowa City: Univ. of Iowa Press.

Smale, L., Holekamp, K. E., Weldele, M., Frank, L., and Glickman, S. E. (1995). Competition and cooperation between litter-mates in the spotted hyena *Crocuta crocuta. Anim. Behav.,* **50,** 671–82.

Smith, C. L. (1965). The patterns of sexuality and the classification of serranid fishes. *American Museum Noviates,* **33,** 1–20.

Smith, C. L. (1975). The evolution of hermaphroditism in fishes. In *Intersexuality in the animal kingdom* (ed. R. Reinboth), pp. 295–310, New York: Springer-Verlag.

Smith, P. S. (1991). Ontogeny and adaptiveness of tail-flagging behavior in white-tailed deer. *Am. Nat.,* **138,** 190–200.

Smith, R. J. F. (1986). Evolution of alarm signals: role of benefits of retaining members or territorial neighbors. *Am. Nat.,* **128,** 604–10.

Smith, R. J. F. (1992). Alarm signals in fishes. *Reviews Fish Biol. Fisheries* **2,** 33–63.

Smith, S. F. (1978). Alarm calls, their origin and use in *Eutamius sonomae. J. Mammalogy,* **59,** 888–93.

Smith, W. J. (1977). *The behavior of communicating: an ethological approach.* Cambridge: Harvard Univ. Press.

Smuts, B. (1985). *Sex and friendship in baboons.* New York: Aldine.

Smuts, B., Cheney, D. L., Seyfarth, R., Wrangham, R., and Struhsaker, T. T. (eds.) (1987). *Primate societies.* Chicago: Univ. of Chicago Press.

Smythe, N. (1977). The function of mammalian alarm advertising: social signals or pursuit invitation. *Am. Nat.,* **111,** 191–94.

Snyder, N. F. R. (1967). An alarm reactions of aquatic gastropods to intraspecific extract. *Cornell Univ., Agri. Experi. Sta., Memoirs,* **403,** 1–122.

Snyder, N. F. R., and Snyder, H. (1970). Alarm response of Diadema antillarum. *Science,* **168,** 276–78.

Snyder, R. J. (1984). Seasonal variation in the diet of the threespined stickleback. *Calif. Fish and Game,* **70,** 167–72.

Soler, M., Soler, J. J., Martinez, J. G., and Møller, A. P. (1995). Magpie host manipulation by great spotted cuckoos: evidence for an avian mafia? *Evol.,* **49,** 770—775.

Solomon, N. G., and French, J. A. (eds.) (1996). *Cooperative breeding in mammals.* Cambridge: Cambridge Univ. Press.

Sordahl, T. A. (1990). The risks of avian mobbing and distraction behavior: an anecdotal review. *Wilson Bull.,* **102,** 349–52.

Soto, C. G., Zhang, J. S. and Shi, Y. H. (1994). Intraspecific cleaning behaviour in *Cyprinus carpio* in aquaria. *J. Fish Biol.,* **44,** 172–74.

Southwick, E. (1982). Metabolic energy of intact honeybee colonies. *Comp. Biochem. Physio.,* **71A,** 277–81.

Southwick, E. (1983). The honeybee cluster as a homeothermic superorganism. *Comp. Biochem. Physio.,* **75A,** 641–45.

Southwick, E., and Mugaas, J. N. (1971). A hypothetical homeotherm: the honeybee colony. *Comp. Biochem. Physio.,* **40A,** 935–44.

Sparks, J. (1967). Allogrooming in primates: a review. In *Primate ethology* (ed. D. Morris), pp. 148–75, London: Weidenfeld and Nicolson.

Stacey, P., and Koenig, W. (eds.) (1990). *Cooperative breeding in birds: long-term studies of ecology and behavior.* Cambridge: Cambridge Univ. Press.

Stammbach, E. (1978). On social differentiation in groups of captive female hamadryas baboons. *Behaviour,* **67,** 322–38.

Stammbach, E. (1988). Group responses to specially skilled individuals in a *Macaca fascicularis* group. *Behaviour,* **107,** 687–705.

Starr, C. (1979). Origin and evolution of insect sociality: a review of modern theory. In *Social insects* (ed. H. R. Hermann), pp. 35–79, New York: Academic Press.

Starr, C. (1985). Enabling mechanisms in the evolution of sociality in the Hymenoptera: the sting's the thing. *Ann. Entom. Soc.*, **78**, 836–40.

Stephens, D. W. (1997a). A diffusion model for a "school" of two fish: directionality, cohesion and tit-for-tat. *Bull. Math. Biol.*, in press.

Stephens, D. W., Anderson, J. P., and Toyer, K. B. (1997b). On the spurious occurence of tit-for-tat in pairs of predator-approaching fish. *Anim. Behav.*, in press.

Stephens, D. W., Nishimura, K., and Toyer, K. B. (1996). Error discounting in the iterated Prisoner's Dilemma. *J. Theo. Biol.*, **167**, 457–69.

Strassmann, J. (1981). Wasp reproduction and kin recognition: reproductive competition and dominance hierarchies among *Polistes annularis* foundresses. *Florida Entom.*, **64**, 74–88.

Strassmann, J. (1989). Altruism and relatedness at colony foundation in social insects. *Trends Ecol. Evol.*, **4**, 371–74.

Strassmann, J., Gastreich, K. R., Queller, D. C., and Hughes, C. (1992). Demographic and genetic evidence for cyclical changes in queen number in a neotropical wasp *Polybia emaciata*. *Am. Nat.*, **140**, 363–72.

Strassmann, J., Hughes, C., Queller, D. C., Turillazzi, S., Cervo, R., Davis, S., and Goodnight, K. (1989). Genetic relatedness in primitively eusocial wasps. *Nature*, **342**, 268–69.

Strassmann, J., and Queller, D. C. (1989). Ecological determinants of social evolution. In *The genetics of social behavior* (ed. M. Breed and R. Page), pp. 81–101, Boulder: Westview Press.

Strassmann, J., Queller, D. C., Solis, C. R., and Hughes, C. (1991). Relatedness and queen number in the neotropical wasp *Paracharergus colobterus*. *Anim. Behav.*, **42**, 461–70.

Struhsaker, T. T. (1967a). Auditory communication among vervet monkeys *(Cercopithecus aethiops)*. In *Social communication among primates* (ed. S. A. Altmann), pp. 281–324, Chicago: Univ. of Chicago Press.

Struhsaker, T. T. (1967b). Behavior of vervet monkeys (*Cercopithecus aethiops*). *Univ. Calif. Pub. Zool.*, **82**, 1–74.

Strum, S. (1984). Why primates use infants. In *Primate paternalism: an evolutionary and comparative view of male investment* (ed. D. M. Taub), pp. 146–85, New York: van Nostrand.

Stubblefield, J. W., and Charnov, E. L. (1986). The origin of eusociality. *Heredity*, **55**, 181–87.

Sturtevant, A. H. (1938). Essays on evolution II. On the effects of selection on social insects. *Q. Rev. Biol.*, **13**, 74–76.

Stutchberry, B. J. (1988). Evidence that bank swallow colonies do not function as information centers. *Condor*, **90**, 953–55.

Stutchberry, B. J., and Robertson, R. J. (1985). Floating populations of female tree swallows. *Auk*, **102**, 651–54.

Stutchberry, B. J., and Robertson, R. J. (1987). Behavioral tactics of subadult female floaters in the tree swallow. *Behav. Ecol. Sociobiol.*, **20**, 413–19.

Sugden, R. (1986). *The economics of rights, cooperation and welfare*. Oxford: Basil Blackwell.

Sulak, K. (1975). Cleaning behaviour in the centrachid fishes *Lepomis macrochirus* and *Microterus salmoides*. *Anim. Behav.*, **23**, 331–34.

Sullivan, K. (1985). Selective alarm calling by downy woodpeckers in mixed-species flocks. *Auk,* **102,** 184–87.

Summers-Smith, J. D. (1963). *The house sparrow.* London: Collins.

Taborsky, M. (1984). Broodcare helpers in the cichlid fish *Lamprologus brichardi:* their costs and benefits. *Anim. Behav.,* **32,** 1236–52.

Taborsky, M. (1985). Breeder-helper conflict in a cichlid fish with broodcare helpers: an experimental analysis. *Behaviour,* **95,** 45–75.

Taborsky, M. (1987). Cooperative behavior in fish: Coalitions, kin groups and reciprocity. In *Animal societies: theories and facts* (ed. Y Ito, J. L. Brown, and J. Kikkawa), pp. 229–39, Tokyo: Japan Scientific Societies Press.

Taborsky, M., and Limberger, D. (1981). Helpers in fish. *Behav. Ecol. Sociobiol.,* **8,** 143–45.

Tamachi, N. (1987). The evolution of alarm calls: an altruism with nonlinear effects. *J. Theo. Biol.,* **127,** 141–53.

Taub, D. (1980). Testing the agonistic buffering hypothesis. I. The dynamics of participation in triadic interaction. *Behav. Ecol. Sociobiol.,* **6,** 187–97.

Tavolga, M. C., and Essapian, F. (1957). The behavior of bottlenose dolphins *Tursiops truncatus;* mating, pregnancy, parturition and mother-infant behavior. *Zoologica,* **42,** 11–31.

Taylor, C. E., and McGuire, M. T. (eds.) (1988). Reciprocal altruism: 15 years later. *Ethol. and Sociobiol.* **9** (special edition).

Taylor, R., Balph, D., and Balph, M. (1990). The evolution of alarm calling: a cost-benefit analysis. *Anim. Behav.,* **39,** 860–68.

Telecki, G. (1973). *The predatory behavior of chimpanzees.* Lewisburg, PA: Bucknell Univ. Press.

Tenaza, R. R., and Tilson, R. L. (1977). Evolution of long-distance alarm calls in Kloss's gibbon. *Nature,* **268,** 233–35.

Terborgh, J., and Wilson Goldizen, A. (1985). On the mating system of the cooperatively breeding saddle-backed tamarin *(Saguinus fuscicollis). Behav. Ecol. Sociobiol.,* **16,** 293–99.

Terry, R. L. (1970). Primate grooming as a tension reduction mechanism. *J. Psychol.,* **76,** 129–36.

Thierry, B., and Anderson, J. A. (1986). Adoption in anthropoid apes. *Int. J. Primat.,* **7,** 191–216.

Thomas, E., and Feldman, M. (1988). Behavior-dependent contexts for repeated plays of the Prisoner's Dilemma. *J. Conf. Resol.,* **32,** 699–726.

Tilson, R. L., and Norton, P. M. (1981). Alarm duetting and pursuit deterrence in an African antelope. *Am. Nat.,* **118,** 455–62.

Tomlinson, J. (1966). The advantage of hermaphroditism and parthenogenesis. *J. Theo. Biol.,* **11,** 54–58.

Toro, M., Abugov, R., Charlesworth, B., and Michod, R. (1982). Exact versus heuristic models of kin selection. *J. Theo. Biol.,* **97,** 699–713.

Toro, M., and Silio, L. (1986). Assortment of encounters in the two-strategy game. *J. Theo. Biol.,* **123,** 193–204.

Trivers, R. (1971). The evolution of reciprocal altruism. *Q. Rev. Biol.* **46,** 35–57.

Trivers, R. (1985). *Social evolution.* Menlo Park, CA: Benjamin Cummings.

Trivers, R., and Hare, H. (1976). Haplo-diploidy and the evolution of the social insects. *Science,* **191,** 249–63.

Trune, D. R., and Slobodchikoff, C. N. (1978). Position of immatures in pallid bat clusters: a case of reciprocal altruism. *J. Mamm.,* **59,** 193–95.

Tschinkel, W., and Howard, D. F. (1983). Colony founding by pleometrosis in the fire ant, *Solenopsis invitius. Behav. Ecol. Sociobiol.*, **12**, 103–13.

Tschinkel, W., and Howard, D. F. (1992). Brood raiding and the population dynamics of founding and incipient colonies of the fire ant, *Solenopsis inviticus. Ecol. Entom.*, **17**, 179–88.

Tsuji, K. (1995a). Inter-colonial selection for the maintenance of cooperative breeding in the ant *Pristomyrmex pungens:* a laboratory experiment. *Behav. Ecol. Sociobiol.*, **35**, 109–13.

Tsuji, K. (1995b). Reproductive conflicts and levels of selection in the ant *Pristomyrmex pungens:* contextual analysis and partitioning covariance. *Am. Nat.*, **146**, 586–607.

Turner, G. F., and Robinson, R. L. (1992). Milinski's tit for tat hypothesis: Do fish preferentially inspect in pairs? *Anim. Behav.*, **43**, 677–79.

Tutin, C. E., and McGinnis, P. R. (1981). Chimpanzee reproduction in the wild. In *Reproductive biology of the great apes* (ed. C. E. Graham), pp. 187–95, New York: Academic Press.

Tuttle, R. (1986). *Apes of the world.* Park Ridge: Noyes.

Uehara, S., Nishida, T., Hamai, M., Hasegawa, T., Hayaki, H., Huffman, M., Kawanaka, K., Kobayashi, S., Mitani, J., Takahata, Y., Takasaki, H., and Tsukahara, T. (1992). Characteristics of predation by the chimpanzees in the Mahale National Park, Tanzania. In *Topics in primatology* (ed. T. Nishida, W. C. McGrew, P. Marler, M. Pickford, and F. de Waal), vol. 1, pp. 143–58, Tokyo: Univ. of Tokyo Press.

Uyenoyama, M. K. (1984). Inbreeding and the evolution of altruism under kin selection. *Evol.*, **38**, 778–95.

Uyenoyama, M., and Bengtsson, B. O. (1981). Toward a genetic theory for the evolution of sex ratio. II. Haplodiploid and diploid models with sibling and parental control of the brood sex ratio and brood size. *Theo. Pop. Biol.*, **20**, 57–79.

Uyenoyama, M. K., and Feldman, M. W. (1980). Theories of kin and group selection: a population genetics perspective. *Theo. Pop. Biol.*, **17**, 380–414.

van Schaik, C.P. (1989). The ecology of social relationships amongst female primates. In *Comparative sociobiology: the behavioral ecology of humans and other animals* (ed. V. Standen and R. A. Foley), pp. 241–69, Oxford: Blackwell Scientific Press.

Vehrencamp, S. (1983). A model for the evolution of despotic versus egalitarian societies. *Anim. Behav.*, **31**, 667–82.

Vieth, W., Curio, E., and Ernst, U. (1980). The adaptive significance of avian mobbing. III. Cultural transmission of enemy recognition in blackbirds: cross species tutoring and properties of learning. *Anim. Behav.*, **28**, 1217–29.

Vine, P. J. (1974). Effects of algal grazing and aggressive behavior of the fishes *Pomacentrus lividus* and *Acanthrus sohal* on coral reef ecology. *Marine Biol.*, **24**, 131–36.

Visscher, P. K., and Seeley, T. (1982). Foraging strategies of honeybee colonies in a temperate forest. *Ecology*, **63**, 1790–1801.

von Frisch, K. (1938). Zur psychologie des Fische-Schwarmes. *Naturwissenschaften*, **26**, 601–6.

von Frisch, K. (1941). Uber einen Schreckstoff der Fischut und seine biologische Bedeutung. *Z. Vergl. Physiol.*, **29**, 46–145.

von Frisch, K. (1967). *The dance language and orientation of bees.* Cambridge, MA: Harvard Univ. Press.

Von Neumann, J., and Morgenstein, O. (1953). *Theory of games and economic behavior.* Princeton, NJ, Princeton Univ. Press.

de Waal, F. (1982). *Chimpanzee politics.* Baltimore, MD: Johns Hopkins Univ. Press.

de Waal, F. B. M. (1989). Food sharing and reciprocal obligations among chimpanzees. *Human Evol.*, **18**, 433–59.

de Waal, F. B. M. (1990). Do rhesus mothers suggest friends to their offspring? *Primates*, **31**, 597–600.

de Waal, F. B. M. (1992). Coalitions as part of reciprocal relations in the Arnhem chimpanzee colony. In *Coalitions and alliances in humans and other animals* (ed. A. Harcourt and F. B. M. de Waal), pp. 233–57, Oxford: Oxford Univ. Press.

de Waal, F. B., and Luttrell, L. M. (1988). Mechanisms of social reciprocity in three primate species: symmetrical relationship characteristics or cognition? *Ethol. Sociobiol.*, **9**, 101–18.

Wade, M. J. (1977). An experimental study of group selection. *Evol.*, **31**, 134–53.

Wade, M. J. (1978). A critical review of the models of group selection. *Q. Rev. Biol.*, **53**, 101–14.

Wade, M. J. (1979). The primary characteristics of Tribolium populations group selected for increased and decreased population size. *Evolution*, **33**, 749–64.

Wade, M. J. (1980). Kin selection: its components. *Science*, **210**, 665–66.

Wallace, A. R. (1891). *Darwinism.* London: Macmillan.

Walther, F. R. (1969). Flight behaviour and avoidance of predators in Thomson's gazelles (*Gazella thomsoni*, Guenther 1884). *Behaviour*, **34**, 184–221.

Waltz, E. C. (1987). A test of the information-centre hypothesis in two colonies of common terns, *Sterna hirundo*. *Anim. Behav.*, **35**, 48–59.

Ward, P., and Zahavi, A. (1973). The importance of certain assemblages of birds as "information centres" for finding food. *Ibis*, **115**, 517–34.

Warner, R. (1984). Mating behavior and hermaphroditism in coral reef fishes. *Am. Sci.*, **72**, 128–36.

Wasser, S. (1982). Reciprocity and the trade-off between associate quality and relatedness. *Am. Nat.*, **119**, 720–31.

Wedekind, C., and Milinski, M. (1996). Human cooperation in the simultaneous and the alternating Prisoner's Dilemma: Pavlov versus Generous Tit-for-Tat. *Proc. Natl. Acad. Sci. USA*, **93**, 2686–89.

Weisbard, C., and Goy, R. W. (1976). Effect of parturition and group composition on competitive drinking order in stumptail macaques *(Macaca arctiodes)*. *Folia Primatol.*, **25**, 95–121.

Weismann, A. (1893). The all-sufficiency of natural selection. *Contemporary Review*, **64**, 309–28.

West-Eberhard, M. J. W. (1969). The social biology of polistine wasps. *Misc. Pub. Museum Zool., Univ. Michigan*, **109**, 1–101.

West-Eberhard, M. J. (1975). The evolution of social behavior by kin selection. *Q. Rev. Biol.*, **50**, 1–35.

West-Eberhard, M. J. (1978). Polygyny and the evolution of social behavior in wasps. *J. Kansas Ento. Soc.*, **51**, 832–56.

Wheeler, J., and Rissing, S. (1975). Natural history of *Veromessor pergandei*. I. The nest. *Pan. Pacific Ento.*, **51**, 205–16.

Wheeler, W. M. (1911). The ant-colony as an organism. *J. Morphology*, **22**, 307–25.

Wheeler, W. M. (1928). *The social insects: their origin and evolution.* London: Kegan Paul, Trench and Trubner.

Whoriskey, F., and FitzGerald, G. (1985). Sex, cannibalism and sticklebacks. *Behav. Ecol. Sociobiol.*, **18**, 15–18.

Wilkinson, G. (1984). Reciprocal food sharing in vampire bats. *Nature*, **308**, 181–84.

Wilkinson, G. (1985a). The social organization of the common vampire bat. I. Patterns and causes of association. *Behav. Ecol. Sociobiol.*, **17**, 111–21.

Wilkinson, G. (1985b). The social organization of the common vampire bat. II. Mating systems, genetic structure and relatedness. *Behav. Ecol. Sociobiol.*, **17**, 123–34.

Wilkinson, G. (1986). Social grooming in the common vampire bat, *Desmodus rotundus. Anim. Behav.*, **34**, 1880–89.

Wilkinson, G. (1987). Reciprocal altruism in bats and other mammals. *Ethol. Sociobiol.*, **9**, 85–100.

Wilkinson, G. S. (1990). Food sharing in vampire bats. *Sci. Am.*, **Feb.**, 76–82.

Wilkinson, G. S. (1992a). Communal nursing in the evening bat, *Nycticeius humeralis. Behav. Ecol. Sociobiol.*, **31**, 225–35.

Wilkinson, G. S. (1992b). Information transfer at evening bat colonies. *Anim. Behav.*, **44**, 501–18.

Williams, G. C. (1966). *Adaptation and natural selection.* Princeton: Princeton Univ. Press.

Williams, G. C. (ed.) (1971). *Group selection.* Chicago: Aldine.

Williams, G. C., and Williams, D. C. (1957). Natural selection of individually harmful social adaptations among sibs with special reference to social insects. *Evol.*, **11**, 32–39.

Wilson, D. S. (1975). A general theory of group selection. *Proc. Natl. Acad. Sci. USA*, **72**, 143–46.

Wilson, D. S. (1976). Evolution on the level of communities. *Science*, **192**, 1358–60.

Wilson, D. S. (1977). Structured demes and the evolution of group-advantageous traits. *Am. Nat.*, **111**, 157–85.

Wilson, D. S. (1980). *The natural selection of populations and communities.* Menlo Park: Benjamin Cummings.

Wilson, D. S. (1983). The group selection controversy: history and current status. *Ann. Rev. Ecol. Syst.*, **14**, 159–87.

Wilson, D. S. (1990a). Weak altruism, strong group selection. *Oikos*, **59**, 135–40.

Wilson, D. S. (1990b). On the relationship between evolutionary and psychological definitions of altruism and selfishness. *Biol. and Phil.* **7**, 61–68.

Wilson, D. S., and Dugatkin, L. A. (1991). Tit for tat vs. nepotism, or, Why should you be nice to your rotten brother? *Evol. Ecol.*, **5**, 291–99.

Wilson, D. S., and Dugatkin, L. A. (1992). Altruism. In *Keywords in evolutionary biology,* (ed. E. Fox Keller and E. A. Lloyd), pp. 28–33, Cambridge, MA: Harvard Univ. Press.

Wilson, D. S., Pollock, G. B., and Dugatkin, L. A. (1992). Can altruism evolve in truly viscous populations? *Evol. Ecol.*, **6**, 331–41.

Wilson, D. S., and Sober, E. (1989). Reviving the superorganism. *J. Theo. Biol.*, **136**, 337–56.

Wilson, D. S., and Sober, E. (1994). Re-introducing group selection to the human behavioral sciences. *Behav. and Brain Sci.*, **17**, 585–654.

Wilson, E. O. (1971). *The insect societies.* Cambridge, MA: Harvard Univ. Press.

Wilson, E. O. (1975). *Sociobiology: the new synthesis.* Cambridge, MA: Harvard Univ. Press.

Wilson, E. O., and Michener, C. D. (1982). Alfred Edward Emerson. *Proc. Natl. Acad. Sci. USA, Biographical Memoirs,* **53**, 159–78.

Witkin, S. R., and Ficken, M. S. (1979). Chickadee alarm calls: does mate investment pay dividends? *Anim. Behav.*, **27**, 1275–76.

Wolf, T., and Schmid-Hempel, P. (1990). On the integration of individual foraging strategies with colony ergonomics in social insects: nectar-collection in honeybees. *Behav. Ecol. Sociobiol.*, **27**, 103–11.

Woodland, D. J., Jaafar, Z., and Knight, M. L. (1980). The "pursuit deterrent" function of alarm calls. *Am. Nat.*, **115**, 748–53.

Woolfendon, G. E., and FitzPatrick, J. W. (1984). *The Florida scrub jay: demography of a cooperative-breeding bird.* Princeton: Princeton Univ. Press.

Wooton, R. J. (1976). *The biology of sticklebacks.* New York: Academic Press.

Wrangham, R. (1975). The behavioral ecology of chimpanzees in Gombe National Park, Tanzania. Ph.D., dissertation. Cambridge University.

Wrangham, R. (1980). An ecological model of female-bonded primate societies. *Behaviour*, **75**, 262–300.

Wrangham, R. (1987). Evolution of social structure. In *Primate societies* (ed. B. B. Smuts, D. L. Cheney, R. M. Seyfarth, R. W. Wrangham, and T. T. Struhsaker), pp. 282–96, Chicago: Univ. of Chicago Press.

Wright, S. (1945). Tempo and mode in evolution: a critical review. *Ecol.*, **26**, 415–19.

Wynne-Edwards, V. C. (1962). *Animal dispersion in relation to social behaviour.* Edinburgh: Oliver and Boyd.

Wynne-Edwards, V. C. (1986). *Evolution through group selection.* Oxford: Blackwell Scientific.

Yahner, R. H. (1980). Barking in a primitive ungulate, *Muntiacus reevesi:* function and adaptiveness. *Am. Nat.*, **116**, 157–77.

Yamamoto, I. (1987). Male parental care in the racoon dog *Nyctereutes procyonoides* during the early rearing stages. In *Animal societies* (ed. Y. Eto, J. L. Brown, and J. Kikkawa), vol. 4, pp. 189–96). Tokyo: Japan Societies Scientific Press.

Ydenberg, R. C., Giraldeau, L. A., and Falls, J. B. (1988). Neighbours, strangers and the asymmetric war of attrition. *Anim. Behav.*, **36**, 433–37.

Zahavi, A. (1975). Mate selection—a selection for a handicap. *J. Theo. Biol.*, **53**, 205–14.

Zentall, T. R. (ed.) (1993). *Animal cognition.* Hillsdale, NJ: Lawrence Erlbaum.

Zuckerman, S. (1932). *The social life of monkeys and apes.* New York: Harcourt Brace.

Author Index

Species Index

General Index